Human Dimensions *of* GLOBAL ENVIRONMENTAL CHANGE

RESEARCH PATHWAYS FOR THE NEXT DECADE

Committee on the Human Dimensions of Global Change
Commission on Behavioral and Social Sciences and Education

Committee on Global Change Research

Board on Sustainable Development
Policy Division

National Research Council

NATIONAL ACADEMY PRESS

Washington, D.C.

NATIONAL ACADEMY PRESS • 2101 Constitution Ave., N.W. • Washington, D.C. 20418

NOTICE: The project that is the subject of this report was approved by the Governing Board of the National Research Council, whose members are drawn from the councils of the National Academy of Sciences, the National Academy of Engineering, and the Institute of Medicine. The members of the committee responsible for the report were chosen for their special competences and with regard for appropriate balance.

The National Academy of Sciences is a private, nonprofit, selfperpetuating society of distinguished scholars engaged in scientific and engineering research, dedicated to the furtherance of science and technology and to their use for the general welfare. Upon the authority of the charter granted to it by the Congress in 1863, the Academy has a mandate that requires it to advise the federal government on scientific and technical matters. Dr. Bruce M. Alberts is president of the National Academy of Sciences.

The National Academy of Engineering was established in 1964, under the charter of the National Academy of Sciences, as a parallel organization of outstanding engineers. It is autonomous in its administration and in the selection of its members, sharing with the National Academy of Sciences the responsibility for advising the federal government. The National Academy of Engineering also sponsors engineering programs aimed at meeting national needs, encourages education and research, and recognizes the superior achievements of engineers. Dr. William A. Wulf is president of the National Academy of Engineering.

The Institute of Medicine was established in 1970 by the National Academy of Sciences to secure the services of eminent members of appropriate professions in the examination of policy matters pertaining to the health of the public. The Institute acts under the responsibility given to the National Academy of Sciences by its congressional charter to be an adviser to the federal government and, upon its own initiative, to identify issues of medical care, research, and education. Dr. Kenneth I. Shine is president of the Institute of Medicine.

The National Research Council was organized by the National Academy of Sciences in 1916 to associate the broad community of science and technology with the Academy's purposes of furthering knowledge and advising the federal government. Functioning in accordance with general policies determined by the Academy, the Council has become the principal operating agency of both the National Academy of Sciences and the National Academy of Engineering in providing services to the government, the public, and the scientific and engineering communities. The Council is administered jointly by both Academies and the Institute of Medicine. Dr. Bruce M. Alberts and Dr. William A. Wulf are chairman and vice chairman, respectively, of the National Research Council.

This study was supported by Contract Nos. 50-DKNA-5-00015 and 50-DKNA-7-90052 between the National Academy of Sciences and the National Oceanographic and Atmospheric Administration. Any opinions, findings, conclusions, or recommendations expressed in this publication are those of the authors and do not necessarily reflect the view of the organizations or agencies that provided support for this project.

Additional copies of the report are available from National Academy Press, 2101 Constitution Avenue, N.W., Lockbox 285, Washington, D.C. 20055; (800) 624-6242 or (202) 334-3313 (in the Washington metropolitan area); Internet www.nap.edu

International Standard Book Number 0-309-06592-5

Printed in the United States of America

iii

iv

Chairman, Committee on Atmospheric Chemistry

WILLIAM L. CHAMEIDES, Georgia Institute of Technology

Chairman, Board on Atmospheric Sciences and Climate

JOHN A. DUTTON, Pennsylvania State University

Chairman, Climate Research Committee

THOMAS R. KARL, National Climatic Data Center, Asheville, North Carolina

Chairman, Committee on the Human Dimensions of Global Change

DIANA M. LIVERMAN, University of Arizona, Tucson

Chairman, Panel on Climate Variability on Decade-to-Century Timescales

DOUG MARTINSON, Columbia University, Palisades, New York

Chairman, Global Energy and Water Cycle Experiment Panel

SOROOSH SOROOSHIAN, University of Arizona, Tucson

Chairman, Global Ocean-Atmosphere-Land System

PETER WEBSTER, University of Colorado, Boulder

NRC Staff

SHERBURNE B. ABBOTT, Executive Director
DAVID M. GOODRICH, Project Director (ending January 16, 1998)
SYLVIA A. EDGERTON, Senior Research Fellow (April 8, 1998, to April 9, 1999)
LAURA SIGMAN, Research Associate (beginning February 17, 1998)
LESLIE McCANT, Project Assistant (beginning January 22, 1999)

Contents

**The table of contents of the entire report
of which this is a part
can be found on the following pages.**

Contents

Global Environmental Change:
Research Pathways for the Next Decade

Preface

This publication is extracted from a much larger report, *Global Environmental Change: Research Pathways for the Next Decade*, which addresses the full range of the scientific issues concerning global environmental change and offers guidance to the scientific effort on these issues in the United States. This volume consists of Chapter 7 of that report, "Human Dimensions of Global Environmental Change," which was written for the report by the Committee on the Human Dimensions of Global Change of the National Research Council (NRC). It provides findings and conclusions on the key scientific questions in human dimensions research, the lessons that have been learned over the past decade, and the research imperatives for global change research funded from the United States.

This publication demonstrates the emergence of a mature research agenda based on an established framework of questions and published findings. It shows how social science provides insights, models, and data of immediate relevance and application to research in earth science (such as projecting carbon emissions or land use change and estimating climate's effects). It also notes progress in understanding of the basic social processes and driving forces underlying the human relationship to the environment (such as public attitudes and population dynamics). In addition, it shows ways in which the social sciences can help direct the priorities of the overall global change program towards more integrated, policy relevant, and effective research imperatives.

Although most of the material on human dimensions research in the larger report appears in this volume, our committee also contributed material on human dimensions to other parts of that report, particularly Chapters 8 and 10, on observations and modeling, respectively. The observations chapter includes a section highlighting the significance of social, economic, and health data to global change research. It notes, for example, the importance of agricultural and population census data in land use research and the need for data on energy production and consumption in research on the carbon cycle. Major challenges include the difficulties of linking social and biophysical data across different scales and spatial units and the lack of data comparability across different political jurisdictions. The chapter also notes that most of the key social data in the United States are collected by agencies, such as the Census Bureau and Department of Health and Human Services, that do not have environmental responsibilities, and such data are not precisely georeferenced. The chapter also addresses issues of confidentiality and privacy raised by detailed human dimensions observations.

The chapter on modeling contains a section that discusses the challenges of including human processes in integrated modeling of the earth system. Noting some progress in integrated modeling, the section highlights the very large uncertainties and great diversity within social systems that limit the predictability of both the human system and the overall earth system. We refer you to these chapters to see the human dimensions issues in context.

The larger report also includes major chapters on ecosystems, seasonal to interannual climate change, decade to century climate change, atmospheric chemistry, and paleoclimate. It makes important recommendations to focus the efforts of the global change research in the United States on central scientific questions and urges the creation of a coherent observational strategy to help answer these questions. We commend the entire report to the social science research and policy communities. Our committee decided to publish Chapter 7 separately in order to reach a community of scholars, students, and policy makers in the United States and elsewhere who have a focused interest in human-environment interactions.

The Committee on the Human Dimensions of Global Change began working a decade ago, at the same time that the U.S. government established the interagency U.S. Global Change Research Program and the U.S. National Science Foundation established a formal program to study the human dimensions of global environmental change. Over that decade, U.S. government support for human dimensions work has expanded modestly, an increasing number of scholars have identified themselves with the field, and efforts have been made to integrate social science into the broader global change programs.

Our committee has offered advice from the scientific community to these efforts within the government. It has worked to develop the intellectual basis for progress in understanding human-environment interactions and to set research directions for the future. The committee's 1992 report, *Global Environmental Change: Understanding the Human Dimensions*, established a framework and research agenda for studying the human causes, consequences, and responses to changes in the global environment such as changes in climate and biodiversity. Since then, the committee has published several reports on specific topics (*Environmentally Significant Consumption: Research Directions* [1997], *People and Pixels: Linking Remote Sensing and Social Science* [1998], and *Making Climate*

Forecasts Matter [1999]) and one report on science priorities—*Science Priorities for the Human Dimensions of Global Change* (1994). However, until the present publication, the committee had not taken the time to reflect on overall progress, the evolution of research questions, and important lessons learned since the 1992 report.

In 1995 the Committee on Global Change Research began a major review of the status of the U.S. research effort on global change. Early in the process, it became clear that human dimensions research was a critical cross-cutting activity across the four themes of the national research program: seasonal to interannual climate prediction, decadal to centennial climate change, atmospheric chemistry, and terrestrial and marine ecosystems. Because our committee had expertise spanning these and other areas, we chose to devote considerable time as a group to developing a chapter for this major review that would identify key science questions, lessons learned, and research imperatives in the field of human dimensions of global change.

I would like to acknowledge the outstanding support and contributions of Paul Stern in helping to prepare the chapter that comprises this report as well as the work and ideas of both present and past members of the Committee on the Human Dimensions of Global Change. Human dimensions researchers Robert Chen, Hadi Dowlatabadi, Greg Knight, Roger Pulwarty and Marvin Waterstone and Robert Costanza of the Committee on Global Change Research also made suggestions of material to be included in the chapter. Very sincere thanks are also due to Berrien Moore and Ed Frieman, chairs of the Committee on Global Change Research and the Board on Sustainable Development, respectively, who encouraged our efforts. I am grateful for the support and encouragement of Barbara Torrey of the NRC's Commission on Behavioral and Social Sciences and Education and Robert Kates of the Board on Sustainble Development, and very appreciative of the assistance and editorial work of David Goodrich, Shere Abbott, Sylvia Edgerton, Laura Sigman, and other NRC staffers who worked so diligently on the report from which this chapter is taken.

Diana M. Liverman, *Chair*
Committee on the Human Dimensions of Global Change (1995-1998)

Human Dimensions of Global Environmental Change:
Research Pathways for the Next Decade[*]

SUMMARY

Research on the human dimensions of global change concerns human activities that alter the Earth's environment, the driving forces of those activities, the consequences of environmental change for societies and economies, and human responses to the experience or expectation of global change. Such research is essential both to understand global change and to inform public policy.

Research on the human causes of global change has shown that socioeconomic uncertainties dominate biophysical uncertainties in climate impacts and possibly also in other impacts of global change. It has shown that human activities, such as deforestation and energy consumption, are determined by population growth, economic and technological development, cultural forces, values and beliefs, institutions and policies, and the interactions among all these things. Ongoing research is improving our understanding of the dynamics of several of these driving forces. It has shown, for example, that human interactions with the environment do not necessarily lead to a "tragedy of the commons" and has begun to enumerate the necessary conditions for successful long-term environmental resource management. Research on the human consequences of global change shows that they are due at least as much to the social systems that produce vulnerability as to environmental changes themselves. This work is refining estimates of impacts and identifying major sources of vulnerability. Research on

---

* This chapter was written by the National Research Council Committee on the Human Dimensions of Global Change, with contributions and editing by the Committee on Global Change Research.

human responses is continually developing and applying better analytical procedures to estimate the costs of global change and policy response options, considering their dependence on highly contestable judgments about nonmarket values and intergenerational equity. Much human response research is focused on the design of human institutions to reduce vulnerability and manage global resources more effectively.

Research over the past decade has made considerable progress, but there are still many unresolved questions. Key research imperatives for the next decade are the following:

- *Understanding the social determinants of environmentally significant consumption.* Research should focus on the most environmentally significant consumption types, changes in consumption patterns as a function of economic growth and development, materials transformations, and the potential for related policy changes. Consumption is a key variable driving trends and patterns in the human impact on atmospheric composition, land use, and biogeochemical cycles.

- *Understanding the sources and processes of technological change.* Research must address the causes of "autonomous" decreases in energy intensity, determinants of the adoption of environmental technologies, and effects of alternative policies on rates of innovation and the role of technology in causing or mitigating global changes.

- *Making climate change assessments and predictions regionally relevant.* Research must develop indicators for vulnerability, project future vulnerability to climatic events, link climate change with social and economic changes in projections of overall regional impacts and their distribution, and improve communication and warning systems, especially in view of recent developments in forecasting.

- *Assessing social and environmental surprises.* The historical record of social and environmental surprises must be explored to clarify the consequences of major surprises, identify human activities that alter their likelihood, and better understand how communication and hazard management systems can help in responding to surprises.

- *Understanding institutions for managing global change.* Research should clarify the conditions favoring institutional success or failure in resource management; the links among international, national, and local institutions; and the potential of various policy instruments, including market-based instruments and property rights institutions, for altering the trajectories of anthropogenic global changes.

- *Understanding land use/land cover dynamics and human migration.* This research should examine and compare case studies of land use and land cover change; develop a typology that links social and economic driving forces to land cover dynamics; and model land use changes at regional and global scales, particularly as they affect ecosystems and biogeochemical cycles and the human consequences of environmental change.

- *Improving methods for decision making about global change.* Research should improve ways of estimating nonmarket values of environmental resources, incorporating these values into national accounts, representing uncertainty to decision makers, and structuring decision-making procedures and techniques of scientific analysis so as to bring formal analyses together with judgments and thus better meet the needs of decision-making participants.
- *Improving the integration of human dimensions research with other global change research.* Human dimensions research supports each of the other fields of scientific research on global change and also addresses key cross-cutting issues. It requires focused and coordinated support that draws on the strengths of both disciplinary and interdisciplinary approaches and that takes advantage of value added by international collaborations.
- *Improving geographic links to existing social, economic, and health data.* Human dimensions data systems benefit from adding geographic information to ongoing social data collection efforts, with appropriate safeguards for confidentiality. The effectiveness of these data systems depends on adequate and stable support. The time is ripe for a careful review of the observational needs for human dimensions research, with careful attention to the ability to link to other observational systems.

INTRODUCTION

Study of the human dimensions of global environmental change encompasses analysis of the human causes of global environmental transformations, the consequences of such changes for societies and economies, and the ways in which people and institutions respond to the changes. It also involves the broader social, political, and economic processes and institutions that frame human interactions with the environment and influence human behavior and decisions. Significant among these are the processes and institutions that use scientific information about environmental processes and human-environment interactions as inputs to human choices that alter the course of those processes and interactions. Thus, one of the human dimensions of global change involves the practical use of scientific information and the issue of how to make such information more useful for decision making. Beginning with a focus on climate change, human dimensions research is expanding to address changes in biodiversity, land and water, pollution, and other globally significant resources and to draw on the extensive literature that addresses human-environment interactions.

Human transformations of the global environment have a long history. Table 7.1 shows that, since 1700, human activity has converted 19 percent of the world's forests and woodlands to cropland and pasture. This shift has altered bio-

TABLE 7.1   Changes in Land Cover, 1700 to 1980

Land Cover Type	Area in 1700 (millions of hectares)	Area in 1980 (millions of hectares)
Forest and woodlands	6,215	5,053
Grassland and pasture	6,860	6,788
Croplands	265	1,501

SOURCE: Adapted from Richards (1990). Courtesy of Cambridge University Press.

geochemical cycles, land surface characteristics, and ecosystems so much that the Earth system itself has changed significantly.

Human activity, especially fossil fuel consumption since the Industrial Revolution, is also responsible for substantial increases in atmospheric concentrations of such gases as carbon dioxide and methane. These increases (see Table 7.2) are mostly associated with the per capita consumption of fossil fuels and growth of the human population; deforestation and the production of cement, livestock, and rice for human consumption; the disposal of wastes from human settlement in landfills; and increased use of fertilizers and industrial and agricultural chemicals. The likely consequences of these gas emissions include a warming of the global climate and a reduction in stratospheric ozone.

Such human activities have accelerated rapidly in recent decades. Between 1950 and 2000 the world's population will have increased from 2.5 billion to more than 6 billion people. Total energy consumption increased from 188,000 petajoules annually in 1970 to almost 300,000 petajoules in 1990, and per capita energy consumption increased from about 50 to 57 gigajoules.[1] Between 1970 and 1990, global forest area decreased by 6 percent, irrigated area increased by almost 40 percent, number of cattle increased by 25 percent, and use of chemical fertilizers doubled.[2]

TABLE 7.2   Greenhouse Gas Concentrations, Preindustrial Age to 1984

Greenhouse Gas	Preindustrial Age	1994	1990s Rate of Change per Year (%)
CO_2	280 ppmv	358 ppm	0.4
CH_4	700 ppbv	1,720 ppb	0.6
N_2O	275 ppbv	312 ppb	0.25
CFC_{11}	0 ppt	268 ppt	0 (HCFC 5%)

NOTES: ppmv, parts per million (volume); ppbv, parts per billion (volume); pptv, parts per trillion (volume).

SOURCE: Intergovernmental Panel on Climate Change (1996b). Courtesy of the IPCC.

These changes, which have altered global environmental parameters, are also associated with improved quality of life for many people: average life expectancy has increased 40 percent since 1955—from 47.5 years then to 65 years in 1995—and infant mortality decreased 60 percent—from 155 deaths per 1,000 in 1955 to 60 in 1990.[3] The rising global averages of per capita energy use, life expectancy, and infant mortality subsume vast disparities. People do not all contribute equally to global change nor benefit equally from progress. The processes determining these changes, sometimes called driving forces, also differ substantially across regions and populations, affecting future trends in both environmental quality and human well-being.

Regional differences in rates of environmental transformation reflect variations in the human driving forces of global change. In the case of greenhouse gas emissions the increase in coal production in China from 7,400 to 21,700 petajoules from 1970 to 1990 represents a doubling of per capita energy consumption due to economic development and national policies. In Mexico oil production grew from 980 to 6,046 petajoules over the same period, reflecting a doubling in per capita energy consumption, significant population growth, and national development policy choices to increase the export of oil. A loss of 40 million hectares of forest in Brazil since 1970 has significant implications for tropical biodiversity, as do losses of almost 10 million hectares each in Indonesia, Thailand, and Mexico. These trends in deforestation result from different combinations of population growth, migration, and economic and policy forces.[4] A major focus of human dimensions research is explaining patterns and changes in the rates of environmental transformation in terms of driving forces that act globally, regionally, and at the level of responsible decision makers.

The impacts of global change on societies and economies are expected to increase greatly in the next century. For example, much of the global change that will eventually result from past human activities has yet to occur, and current trends in these activities portend potential large increases in global change. As the major climatic changes lie in the future, so do their implications for humanity. This may also be true for the human consequences of ecological transformations now occurring through deforestation and other anthropogenic land cover changes. Thus, another major focus of human dimensions research is estimating the social and economic consequences of anticipated global environmental changes. This research integrates information about anticipated environmental changes with information on the social parameters that determine the impact of those changes: demand for affected natural resources, vulnerability of geographical regions and social groups to particular environmental changes, and the potential for adaptive response. In addition, human dimensions research addresses the workings of social systems that manage environmental resources—markets, property rights regimes, treaties, legal and informal norms, and so forth—and the potential to modify those institutions through policy and thus to mitigate global change or increase adaptive capability.

In sum, human dimensions research aims at understanding how human activity drives greenhouse gas emissions, regional air quality, land cover change, and alterations in terrestrial and marine ecosystems; predicting the course of the activities that drive those transformations; estimating how changes in climate, land cover, ecosystems, and atmospheric chemistry affect food, water, natural resources, human health, and the economy; analyzing the ways that societies manage environmental resources; and analyzing the feasibility and possible costs and implications of technical, economic, behavioral, and policy responses to those environmental changes. This research builds basic understanding of human-environment interactions and provides information and responsive tools to decision makers.

Although research on the social and policy aspects of environmental change has a long history, human dimensions research only became formally linked to global change research in the late 1980s. The potential for making this link was set forth in seminal writings addressed to national and international research policy makers.[5] Human dimensions research became part of the U.S. Global Change Research Program (USGCRP) in 1989 with a small National Science Foundation (NSF) program and has since become a significant component of the USGCRP. This activity, together with more general support from government, foundations, and universities for social science research on global change, has resulted in some significant accomplishments and insights in understanding the human dimensions of global climate change.

CASE STUDIES: CONTRIBUTIONS OF HUMAN DIMENSIONS RESEARCH IN ADDRESSING GLOBAL CHANGE

Human Dimensions Research and the IPCC

Contributions to the recent Intergovernmental Panel on Climate Change (IPCC) reports are a good illustration of the significance and policy relevance of human dimensions research. The year 1988 is often identified as a turning point in public and political perceptions of climate change in the United States. While the news media linked drought to global warming, scientists, environmental groups, and decision makers gathered in Toronto to declare the need for a 20 percent cut in greenhouse gas emissions.[6] Meanwhile, social and applied scientists were working to develop methods for assessing the economic and social consequences of climate change and examining the implications of the policies that might be used to mitigate it. The results of this research were reported to the IPCC and became an important part of international debate and decision making in response to the threat of climate change.

For example, demographers, geographers, and others have estimated populations at risk from sea level rise and demonstrated the tremendous vulnerability of many large cities to climatic variations.[7] The synthesized results of many country case studies indicated many billions of U.S. dollars in potential losses and protection costs associated with a 1-meter rise in sea level (see Table 7.3).

TABLE 7.3   Impacts of a 1-Meter Sea Level Rise in Selected Countries

Country	People Affected (millions)	Economic Loss (billions of U.S. dollars)	Land Area Lost (km²)	Protection Cost (billions of U.S. dollars)
Bangladesh	71	NA	25,000	1+
China	72	NA	35,000	NA
Egypt	4.7	59	5,800	13.1
Japan	15.4	849	2,300	156
Netherlands	10	186	2,165	12.3
United States	NA	NA	31,600	156

NOTE: NA, not available.

SOURCE: Bijlsma (1996). Courtesy of the Intergovernmental Panel on Climate Change (IPCC).

To estimate the potential effects of global warming on the world's food system, agronomists and economists linked the output of climate models to crop yield and economic models.[8] Figure 7.1 shows several important results of these studies, including the sensitivity of impact assessments to the results of different climate models, the considerable potential for adaptation to alter the impact of climate change, and the relative vulnerability of developing countries.

Also important to the IPCC and other assessments are efforts to calculate the costs and benefits of various mitigation strategies, such as carbon taxes and carbon sequestration through reforestation, including estimates of nonmarket values. For example, the estimated costs of a carbon tax to achieve a 20 percent reduction in CO_2 emissions ranged from $50 to $330 per ton of carbon in the IPCC study,[9] depending on the economic assumptions and model used. Forest plantations and forest management have the potential to sequester up to 75 billion tons of carbon a year.[10] Studies of the economic feasibility of this strategy have been used as a basis for discussions in the negotiations for the Framework Convention on Climate Change and have informed debate on strategies such as joint implementation of carbon reductions through aid for forest and energy efficiency projects.

Also considered by the IPCC was the issue of deforestation in Amazonia, where human dimensions research has informed policy decisions in Amazonian nations, especially Brazil, and in international organizations such as the World Bank. In the late 1980s international attention focused on Amazonia, where rapid deforestation was linked to climate change, loss of biodiversity, and threats to indigenous peoples.[11] Human dimensions research revealed the causes of forest destruction; for example, the building of highways opened the forest to migrants, many of whom did not know how to farm cleared land or manage forests sustainably.[12] Biases in agricultural subsidies, tax incentives, and high inflation promoted extensive land clearing for ranching.[13] Detailed social and spatial analyses of relationships among deforestation, secondary growth, and demo-

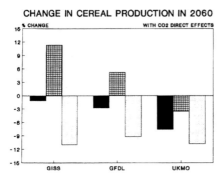

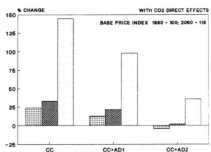

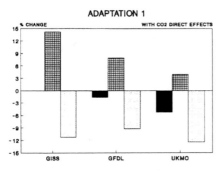

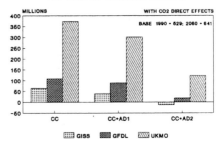

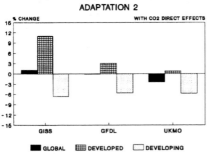

■ GLOBAL ▦ DEVELOPED □ DEVELOPING

FIGURE 7.1 Change in global, developed country, and developing country cereal production, cereal prices, and people at risk of hunger in 2060 under different climate change scenarios (% change from a base estimate for 2060). NOTES: GISS, Goddard Institute of Space Science; GFDL, Geophysical Fluid Dynamics Laboratory; UKMO, U.K. Meteorological Office; CC, climate change scenario including direct CO_2 effects; Adaptation 1 (AD1), adaptation level involving minor changes to existing agricultural systems; Adaptation 2 (AD2), adaptation level involving major changes. Reference scenario assumes no climate change. SOURCE: Rosenzweig and Parry (1994). Courtesy of Macmillan Magazines Ltd.

graphic characteristics showed heterogeneous patterns that challenged simple explanations of land use change and showed the need for local strategies for ecosystem protection.[14]

Partly as a result of these research insights, countries such as Brazil have altered taxation and subsidy structures that favored ranching and have adopted policies for more sustainable development of forest lands. Multilateral development agencies now undertake environmental assessments for transportation and other development projects. Popular accounts are now more sensitive to the varied causes and responses to Amazonian deforestation.

Consequences of Climate Change and Variability at the Regional Level

Researchers have compiled data on overall losses from climatic disasters and have shown that economic damages are increasing dramatically, especially in the United States. For example, hurricane and flood losses have reached more than $1 billion annually in recent years and have stressed both federal disaster relief and private insurance systems.[15] Although these increased disaster losses may be due to climate change, much of the increase is a result of increasing vulnerability resulting from more people living in hazard-prone locations, increasing property prices, and inadequate land use and building regulations. In the developing world, millions of people have been displaced by cyclones, flooding, and droughts, as population growth, migration, and poverty expose more people to climatic extremes.[16] The human consequences of climate change and variability depend critically on the vulnerability of human populations and on their ability to adapt, as well as on climatic events.

Studies have also identified a serious threat of changes in the patterns of diseases and pests associated with climate change and variability.[17] The 1993 Midwest floods were associated with multiple epidemics in the United States. Heavy rains in Milwaukee overwhelmed the sanitation system, creating a plume of farm waste and contaminated runoff in Lake Michigan that later entered the water supply, resulting in a large outbreak of *Cryptosporidium* (400,000 cases, with more than 100 deaths). In Queens, New York, an exceptionally hot, humid summer boosted local mosquito populations, leading to local transmission of malaria. In the southwestern United States, intense rains provided a sudden burst of food supplies for rodents, following a six-year drought that significantly reduced rodent predators (owls, coyotes, and snakes). The 10-fold rise in rodents led to transmission of a "new" disease—hantavirus pulmonary syndrome—with a case fatality rate of 50 percent.

In Southern Africa, prolonged drought, punctuated by heavy rains in 1994, precipitated an upsurge of rodents, crippling agricultural yields in Zimbabwe and leading to plague in Mozambique and Malawi. In India in 1994 flooding following a summer of 51°C temperatures across the plains led to an outbreak of rodent-borne plague, as houses with stored grains heated up, generating clouds of fleas. In addition to severe human losses in the affected regions, measured economic losses included $2 billion to $5 billion by international airline and hotel chains from lost tourism.

Extreme weather events compounding local vulnerabilities (multiple stresses) can disrupt predator/prey relationships (functional biological diversity) and can generate biological surprises, such as population explosions of pests and pathogens that can affect human, plant, and animal health. The impacts of extreme events and epidemics can ripple through economies, affecting agriculture, productivity, trade, and tourism, in addition to their direct effects on regional human health and well-being.

There is, of course, much uncertainty about the role of climatic change in causing ecological changes that have costly effects on humans. A major recent example that highlights the difficulty in assigning causation is the collapse of the commercially important northern cod populations off the coast of eastern Newfoundland and Labrador, Canada, in the late 1980s and early 1990s. This collapse led to a costly program to compensate the over 30,000 people who could no longer work as fishers or fish processors. Debates continue about the roles of the North Atlantic Oscillation and other, more specific, climatic and oceanographic changes relative to the role of overfishing.[18] There is also uncertainty about the links from ecological consequences to human consequences because of gaps in knowledge about the ability of human communities to respond effectively to anticipated ecological changes.

In those regions where climatic variability is associated with El Niño-Southern Oscillation (ENSO) events, there is hope that improved understanding of sea surface temperatures and associated changes in atmospheric circulation will result in advance warnings of droughts, floods, and epidemics and reduced losses.[19] This type of human dimensions research highlights the importance of improved understanding of climate change and variability, the need to consider social vulnerability and adaptive capacity when forecasting the consequences of global change, the potential benefits of predicting climatic extremes, and the need to evaluate carefully options for reducing greenhouse gas emissions.

KEY SCIENTIFIC QUESTIONS

Key scientific questions for research on the human dimensions of global change can be grouped into four broad interrelated interdisciplinary categories:

- What are the major human causes of changes in the global environment and how do they vary over time, across space, and between economic sectors and social groups?
- What are the human consequences of global environmental change for key life support systems, such as water, health, and agriculture, and for economies and political systems?
- What are the potential human responses to global change? How effective

are they and at what cost? How do we value and decide among the range of options?

- What are the underlying social processes or driving forces behind the human relationship to the global environment, such as human attitudes and behavior, population dynamics, and economic transformation? How do they function to alter the global environment?

Research on the human dimensions of global change has value both as basic science and for informing environmental decisions. It increases basic understanding of how past human activities have created present environmental conditions, how past environmental changes and variations have affected human well-being, and how people have responded to these variations and changes. By developing understanding of human-environment dynamics, human dimensions research improves the knowledge base for anticipating future environmental changes and for informing policies aimed at reshaping the environmental future. Studies of the human consequences of and responses to global change help inform judgments about what kinds of responses would be most desirable (e.g., mitigation, adaptation options) and about how to organize those responses to achieve the desired effects. Below we describe the major science issues, review progress that has been made in understanding them, and identify some lessons that have been learned from previous research.

What Are the Major Human Causes of Changes in the Global Environment?

What has been learned in recent years about human causes of global environmental change? One major focus of research has been the explanation of changes in the composition of the Earth's atmosphere. Looking at the atmosphere through human history, one finds that the concentrations of several gases (carbon dioxide, methane, nitrous oxide) changed only a little for more than a thousand years and then started to increase rapidly around 1800. The obvious hypothesis to explain these data is that prior to industrialization in the nineteenth century the related basic cycles of the Earth's environment were in approximate equilibrium and aggregate human activity was too small to be detectable in globally averaged data; then, increasingly since the Industrial Revolution, aggregate human activity has changed the composition of the atmosphere, in particular adding measurably to the concentrations of certain gases. Similarly, looking at the history of land use and land cover, one finds significant changes occurring, although over longer time periods. The obvious hypothesis to explain these observations again is that human beings altered the land and used resources to meet the needs of a rapidly growing population and an expanding industrial economy. Research into the direct human causes of global change has thus focused on changes in land and energy use. But there is also a growing body of work on the fundamental social processes that drive human use of the environment.

Human Activity and Land Use Change

Interest in the causes of local and regional land use changes is long standing in the social sciences.[20] Significant steps have been made in documenting the long history of human transformation of land cover and in explaining the major forces that drive land use. These studies are of interest to a wide range of social and environmental scientists because land is a key factor in social relationships and resource use. But these studies also provide specific contributions to scientific understanding of biogeochemical cycles (especially the carbon cycle), regional climate modification, and alterations in natural ecosystems and are a critical basis for policies to mitigate and adapt to climate change, conserve biodiversity, and reduce land degradation.[21] Land use studies provide a powerful rationale for maintaining land and marine remote sensing satellite systems and suggest ways in which these technologies can be made more germane to decision making.

The global change research community has made considerable progress in recent years on several important questions, such as the social causes of deforestation in regions like the Amazon River basin and Southeast Asia; the role of social, political, and economic institutions in land use decisions; and the relationships between population and land use (and land cover) change.[22] There have also been tremendous improvements in the ability to combine social, physical, and remote sensing data within geographic information systems, often with the explicit purpose of understanding how processes at local scales are nested in regional, national, and global scales.[23]

Additionally, human dimensions research has highlighted the important distinction between land use and land cover. Whereas *land cover* refers to the land's physical attributes (e.g., forest, grassland), *land use* expresses the way such attributes have been transformed by human action (e.g., ranching, crop production, logging); that is, land use measures provide a socioeconomic portrait of a landscape.[24] Land cover is directly represented in global climate models. Land use links land cover to the human activities that transform the land.

The emerging field of environmental history has provided important data on the trajectories and causes of land use changes in the past. For example, historical studies of the U.S. Great Plains have shown how changes in the use and management of grazing and croplands relate to government policy and economics and in turn influence the cycling of carbon and nutrients.[25] Historians and geographers have also reconstructed the history of human use of such regions as the Mediterranean, Caribbean, and Latin America.[26]

Historical studies of land use have altered scientific thinking on the past and the present in a variety of ways. For instance, many observers have presumed that much of the humid tropical forests is pristine or that human impacts on the global environment mainly occurred in recent decades. However, research has shown that many forests were cleared in the distant past or have been managed for centuries and that their current rich biodiversity may be a product of past human manipulation, resulting in higher frequencies of species with economic, medicinal, and other human uses than might be expected to result from natural processes of secondary succession.[27]

Although human population growth is commonly seen as the major cause of land cover change and destruction of habitats for biota, particularly because of land clearing to grow food, the role of population is in fact far more complex. Numerous cases do suggest that population growth and/or migration are correlated with increasing rates of tropical deforestation, but just as many suggest that population growth need not lead to increasing deforestation—when alternative employment, settlement concentration, and other processes are available as alternatives to land clearing, to provide a population with an acceptable standard of living.[28] In fact, there is considerable evidence that only at higher population densities does one find more intensive and efficient use of land.[29]

Research on land management practices has demonstrated that over-exploitation of common-pool natural resources—the so-called tragedy of the commons[30]—is not an inevitable consequence of human nature and the spatial distribution of resources but is contingent on the structure of human communities and the condition of social institutions that effectively govern access to a resource, monitor its condition, and establish sanctions for overexploitation.[31] Both cultural traditions and contemporary legal rules, such as land tenure rules, are critical in influencing how land can be used and by whom.

The emergence of integrated and interdisciplinary approaches to understanding land use and environmental issues has resulted in a series of studies that show how political and economic structures constrain individual choices about management of land and resources.[32] For example, colonial legacies of unequal land tenure and export-oriented production, combined with current unfavorable terms of trade and debt, have driven many peasants to overuse their land, adopt polluting technologies, or cut their forests.[33]

Social scientists have begun to make greater use of orbital Earth-observing satellites in recent years. The interest in understanding the social dimensions of land use change has challenged some of the inferences about land use drawn by natural scientists by showing, for example, the importance of secondary growth and the likely miscalculations of biomass and carbon pools resulting from overly aggregated analyses that fail to quantify the differences between mature and 10-year-old regrowth vegetation.[34] Social scientists have explained the processes underlying various patterns of forest change seen on satellite images in terms of the development of transportation networks, land tenure, and export agriculture. Social scientists have also made important contributions to explaining satellite observations of vegetation dynamics in Africa and to understanding land use change in areas undergoing urbanization.[35]

Field studies of land use have provided information of great relevance to global and regional atmosphere-biosphere modeling. For example, coarse-resolution satellite data tend to represent the predominant soil type or vegetation in each grid cell, even if a minor soil or vegetation type is of major economic or ecological significance. Such a representation of the data can seriously misrepresent land use and productivity potential as well as biogeochemical cycles. Another important development is the focus on explaining trends and patterns in land use intensification, in which crop yields are increased through the use of agricultural chemicals and irrigation, resulting in alterations in regional and global biogeochemical cycles and ecosystems.

Progress in the past decade is evident in the rise of an International Human Dimensions Programme/International Geosphere-Biosphere Programme (IHDP/IGBP) core project on land use/land cover change, with a coordinated, comparative, multilevel strategy for understanding, monitoring, and modeling land use.[36] In developing frameworks, case studies, and models of how social forces drive changes in land use and land cover, this type of comparative research program has the potential to explain and predict land use change but also to assist in identifying strategies for managing land use and protecting ecosystems.

Recent important U.S. initiatives include the expansion of the population program at the National Institute of Child Health and Human Development (NICHD) into population and environmental research in 1995, the creation in 1996 of an NSF-funded research center that works on land use—the Center for the Study of Institutions, Population and Environmental Change at Indiana University—and the National Aeronautics and Space Administration's (NASA) Land Use Cover Change request for proposals. In summary, there has been considerable progress in understanding the human causes of land use change, including the following insights:

- Humans have been altering land cover and use for centuries.
- Some regions that now appear pristine have been subject to human management since prehistoric times.
- There is no simple relationship between population and deforestation or between common property rights and resource degradation.
- The analysis of institutions—in their broadest sense, including political, legal, economic, and traditional institutions—and their interactions with individual decision making is critical in explaining land use.
- Satellite images can provide important insights for social science, and social science can help guide satellite programs to useful applications.
- The age and gender structure of landholding households affects how much forest is cut for farming.
- Tax incentives affect Amazonian deforestation.[37]
- Secure land tenure is important to long-term resource conservation.[38]

- Road construction in forests leads to increased deforestation not only by farmers claiming land but also by logging companies.

However, there is still inadequate knowledge on such key issues as these:

- How to develop land management institutions that both respond to local needs and mitigate global environmental change.
- How to aggregate in-depth studies of land cover and land use to provide global projections of use in large-scale modeling and international management of global change.
- The role of population mobility in land use change.
- How to best use the expanding range of satellite data in land use/land cover change research.

Human Impacts on Coastal and Marine Ecosystems

Global change research encompasses the study of changes in coastal and marine ecosystems insofar as they are affected by physical and socioeconomic processes that are global in scale and effect. Social and applied scientists have investigated the importance of coastal and marine ecosystems for many communities, regions, and nations. They have also addressed the ways in which resource use and pollution have altered the condition and biodiversity of coastal ecosystems in many regions of the world, including the destruction of protective and productive mangrove ecosystems, the degradation of coastal lagoons and estuaries and species that live or reproduce in them, and the minor contamination of even the deep and remote oceans.

Steady increases in demand, technological capacity, and effort have led to a long-term trend of increasing fish catches, which is believed to have leveled off during the 1990s, indicating limits to sustainable harvests.[39] Heavy fish mortality means that environmental fluctuations as well as other human impacts, such as pollution and degradation of habitat, make fisheries even more vulnerable.[40] Social scientists and others have documented the roles of technological change, population growth, institutional structures, and social attitudes in driving demand for fish and other marine resources, as well as in shaping the nature and effectiveness of fisheries management, and they have sought ways to use these resources more sustainably.[41] They have also contributed to understanding the ecological and social concerns associated with mariculture, which is increasing throughout the world as a way to compensate for declining natural resources.[42] This research also contributes to several related themes identified in this chapter, including the links between economic globalization (e.g., of industrial shrimp farming), conflicts over common property resources and loss of forest lands (mangroves); the emergence of new social institutions (social movements in resistance to industrial aquaculture); and the use of new information technologies (communications and spatial) in resource management.[43] Research on common property management, discussed in a later section of this chapter, has drawn many important examples from marine ecosystem use.[44]

Some important insights of this research include the following:

- People have responded to problems in coastal marine systems primarily by intensifying, diversifying, and expanding the areal extent of their uses of those systems, tending to extend such problems to the global level.
- Globalized systems of production and marketing, combined with increases in population and consumer demand and patterns of subsidization, increase competition between countries and communities for scarce marine resources.
- Rules of free and open access, combined with the weaknesses of international management regimes, make it difficult to control harvesting in deep ocean and other multinational fisheries.
- Restricting access is a necessary but not sufficient approach to reducing incentives to overharvest and pollute marine ecosystems.
- The technical and institutional tools of marine resource management have not adequately incorporated the effects of coastal development, wetlands drainage, dams, and pollution of rivers and oceans in diminishing breeding habitat and degrading marine resources.
- The success of fisheries and coastal management depends on functional interdependence between local institutions and regional, national, and international institutions.

Current knowledge is not adequate to achieve several essential goals:

- Provide complete geographic coverage of the status of human use of marine and coastal resources.
- Analyze and model changes in the abundance of fish and marine mammal populations as a function of multiple social and environmental stresses, including interannual, decadal, and longer-term climatic change.
- Evaluate the full range of institutions, including traditional systems, to understand how they increase or reduce human impacts on coasts and oceans.

Changes in Energy and Materials Use

Fossil fuel use is the most prominent human activity that alters the composition of the global atmosphere. Since the 1970s, a burst of human dimensions research seeking to understand the consumption of fossil fuels has been proceeding simultaneously at several levels. The methods developed for studying energy use have more recently been applied to human transformations of the global nitrogen cycle and to human consumption of other environmentally significant materials. The results are useful as inputs to climate models, for anticipating future rates of environmental change, and for identifying effective ways to mobilize social and economic forces to alter trajectories of environmental change.

First, fossil fuel use has been disaggregated by fuel type, geographic region, mediating technology, and social purpose (lighting, water heating, transportation, steel making, etc.). It has been shown that patterns and environmental impacts vary greatly by country and that national-level consumption varies with technology, population, and other factors, such as industrialization and degree of central planning of economies.[45]

Second, progress has been made in understanding patterns and changes in energy and materials use across countries and over time.[46] Energy use and its environmental impacts, for example, generally increase as a function of economic prosperity, but there are exceptions. Countries beyond a certain level of affluence experience declines in per capita environmental impact, although considerable dispute remains about where the turning point lies.[47] Also, the energy-affluence link breaks down in certain periods, including those characterized by rapidly increasing relative energy prices and significant policy interventions.[48] Thus, changes in prices and policies allowed economic growth to continue in the United States without increases in energy use or carbon emissions between the mid-1970s and the mid-1980s, but energy use has been increasing since then, driven by increasing travel demand, shifts in the vehicle fleet, and other factors, and similar trends have been occurring in other developed countries.[49] Long-term trends show decreasing carbon emissions per unit of energy use due to fuel switching and electrification, decreasing materials use per unit of economic output, and replacement of dense materials such as steel with lighter-weight materials such as plastics.[50] These rates of change have an autonomous dynamic and respond to the prices of inputs, but little is known about how public policy might alter the trends to enhance environmental quality.

Third, patterns of energy and materials use have been studied in relation to particular variables that may account for changes and variations in use, and some of these variables can be affected by public policy. At the household level, for example, energy use is affected by income and fuel prices, household structure and social group membership, and by individual knowledge, beliefs, and habits, as well as by the energy-using technologies that households possess.[51] Research on the determinants of consumer decisions to take advantage of technical and economic possibilities to improve energy efficiency indicates that more is required than favorable attitudes and accurate information. There is significant potential to improve residential energy efficiency with appropriately designed interventions. The research strongly suggests that the most effective interventions are specific to consumers' situations and that they use combinations of information, incentives, and social influence. Participation of affected consumers in program design can greatly increase effectiveness.[52]

Research has focused on identifying how the energy consumption patterns of firms and individuals change as a function of changes in information, incentives, technology, and social organization, thus illuminating the potential for reducing society's reliance on fossil fuels by promoting the adoption of new technologies or changing behaviors and preferences. Specific areas of extensive research include technology-focused research on energy consumption, energy efficiency, renewable energy, and nuclear power; research on price elasticities and response to incentives; and research on behavioral and informational factors affecting change in consumer choice.[53] Consumer energy choices are also shaped by political and economic structures that influence regulations and incentives for different types of energy, transportation, and housing policy, as well as the reach of advertising to different regions and social groups.

Research on energy conservation has blended behavioral and technological analyses to compare the technical potential for reducing the energy use required to provide an energy service, such as indoor climate control, with actual reductions in energy consumption. It has examined ways to achieve more of this potential reduction by identifying and removing barriers to energy conservation, such as subsidies and other market distortions, principal-agent problems, incomplete consumer knowledge and misinformation, and problems related to the early stages of the introduction of new technology. This research provides a basis for selecting promising policy options to achieve national commitments to stabilize greenhouse gas emissions.

Materials balance analysis provides the basis of an accounting system that tracks the stocks and flows of certain materials, particularly the chemical elements, through the human economy. Analysis of material flows in this fashion has been called industrial metabolism and industrial ecology.[54] As in energy research, the analysis begins with descriptive work that clarifies the principal human activities dominating each materials flow; proceeds to explorations of behavioral, economic, technological, and policy-related determinants of these activities; and expands to include prospects for changes in consumption patterns over time, including changes related to economic development.[55] One element that has been productively studied is nitrogen, which in the form of nitrous oxide acts as a greenhouse gas and affects the chemistry of stratospheric ozone and in several chemical forms plays a role in nitrogen fertilization of the biosphere. The predominant role of fertilizer production in human-induced changes to the global nitrogen cycle was only recently recognized.[56]

Our understanding of energy use is far more sophisticated than it was two decades ago. It has led to the broader concept of environmentally significant consumption and to the idea of applying analyses like those used for energy to various nonelemental materials of environmental importance, such as wood, steel, cement, glass, and plastics.[57]

An important part of research on energy use is scenario making, which seeks to extrapolate current energy use patterns into the future.[58] Time horizons of prominent scenarios range from a decade to a century. Two important goals of scenario making are transparency (explicit reviewable assumptions) and self-consistency. For global change studies, most scenarios are built on the basis of models of the evolution of national economies, often assuming a similar evolution for large groups of countries at a similar stage of economic development and structure. Typically, population growth is exogenous to the models, and per capita energy consumption is the subject of investigation. The most significant uncertainties relate to the determinants of the rate of introduction of new technologies.[59] Rarely addressed to date but presenting major sources of uncertainty are changes in preferences over time, such as those that might accompany new environmental information.[60]

Scenario making is a conservative activity in that it assumes only slow changes from established trends; it is not well suited to exploring the significance of surprises and catastrophes. Nonetheless, over the past two decades, scenario making to elucidate energy consumption has become a highly developed art, featuring dialogues among modelers to ensure quality control and inter-comparisons and to highlight debatable assumptions. Scenario building has been an essential basis for IPCC assessment models of future climate and analyses of mitigation options, the latter employing models used for scenario building in policy analysis of greenhouse gas emission control strategies.[61]

Important insights of such activities include the following:

- Estimates of future emissions of greenhouse gases are highly sensitive to assumptions about future economic, technological, and social changes, particularly about the autonomous rates of decarbonization and improvement in the energy efficiency of technology, about the likelihood of further large-scale economic transformations, and about the stability of preferences.
- Energy and materials uses are determined by multiple factors: they are not simple functions of population or economic activity but depend on complex interactions of these factors and others.

Future emissions of greenhouse gases will be driven by pressures from increasing affluence and population, with countervailing trends that reduce the amount of energy and materials used per unit of economic activity and the rate of emissions per unit of energy and materials used.

Current knowledge is inadequate to accomplish some tasks critical to understanding consumption trends, their potential environmental consequences, and the possibilities for altering them. These tasks include:

- Clarifying the determinants of "autonomous" change in energy and materials efficiency and thus improving the accuracy of projections of change in greenhouse gas emissions and in the pressure on depletable resources.

- Specifying the ways in which population, technology, affluence, preferences, policies, and other forces interact to change the rates of environmentally significant consumption in high-consuming developed economies and particularly in developing economies, where large increases in consumption are anticipated.
- Identifying and quantifying important sources of variation in the adoption of environmentally beneficial technology among firms within industries.

What Are the Human Consequences of Global Environmental Change?

Human dimensions research has made important progress in understanding the consequences of global change for people and ecosystems. Drawing on earlier research in applied climatology and natural hazards, the past 10 years have seen a major effort to understand the potential impacts of climate change on human activity, as well as studies of the impacts of past and present climate variability, the impacts of ozone depletion on human health, and the effects of land degradation and biodiversity loss on society. Credible climate impact assessments are a basis for developing policy responses to global climate change and for successful application of information on current climate variability to resource management.

Consequences of Future Climate Changes

The first studies of potential global warming impacts analyzed how crop yields and water resources would change in developed countries in response to climate scenarios of monthly changes in temperature and precipitation, based on coarse and uncertain output from climate models simulating the equilibrium response to a doubling of carbon dioxide levels in the atmosphere.[62] Later crop modeling efforts have incorporated the direct physiological effects of higher carbon dioxide levels, employed transient climate scenarios and daily data, covered developing countries, and replaced the concept of the unresponsive farmer with that of people capable of flexible adaptation to climate change.[63]

It appears that many U.S. farmers will be able to adapt to the climate changes expected from a doubling of atmospheric carbon dioxide levels by shifts in technology and crop mix but that others, especially in developing countries, will experience lower yields because they cannot afford technology and may be farming more biophysically vulnerable land.[64] Some studies of economy-wide impacts arrive at similar conclusions;[65] however, these conclusions may be sensitive to some of the assumptions underlying the analyses, as discussed in more detail in the section below on integrated assessment.

A major conceptual advance occurred in moving from impact assessments based on climate model scenarios to analyses based on an understanding of vulnerability.[66] The lack of consensus about how climate may change at the regional level, and the recognition that changes in social systems may be more important than changes in natural systems in determining the impacts of drought and other climatic shifts, reoriented the Working Group 2 of the second IPCC assessment to pay more attention to vulnerability assessment. For example, rapid increases in water demand have increased drought vulnerability, and the spread of urban settlements into coastal and flood-prone regions has increased vulnerability to sea level rise and severe storms.

Another innovation is the development of multisectoral regional assessments of the consequences of climate change. The Missouri-Illinois-Nebraska-Kansas (MINK) and Mackenzie River basin studies are good examples of a regional and cross-sectoral approach to climate impact assessment. The MINK study examined what would happen if the drought conditions of the 1930s were imposed on the economy and resources of the MINK region of the future. Taking into account adaptation, the study showed that, on the whole, agriculture would be able to cope with climate change better than forests or water resources and that impacts on the regional economy would be minor.[67] The Mackenzie River basin study examined the impacts of climate change in the Canadian Arctic using several climate and socioeconomic scenarios and including local stakeholders in defining policy questions and potential responses. The study found significant effects on northern ecosystems and hydrology.[68]

These new approaches to climate impact assessment have relevance far beyond the study of global warming. Many of the methods can be used to understand the impacts of seasonal to interannual climate variability, thus increasing the usefulness of forecasts on that timescale based on improved understanding and modeling of El Niño. The new approaches can also be used in analyzing the impacts of decadal shifts in atmospheric circulation and climate variability. For example, statistical crop models were used[69] to correlate ENSO with maize yields in Zimbabwe, and several early warning systems for famine also drive crop models with climate information to manage food security.

One of the most significant emerging areas of research is the effects of climate change and variability on human health. Several studies have shown that climate change may result in changes in the incidence and prevalence of such diseases as cholera and malaria, which debilitate human populations,[70] as well as changes in the geography of crop and livestock pests and diseases. For example, as plants have migrated upslope in the highland tropics in response to warmer climate, and tropical summit glaciers have generally retreated, and freezing levels in the mountains have moved up 150 m since 1970, mosquito-borne diseases have increasingly infested highlands and high-elevation cities such as San José, Costa Rica, and Nairobi, Kenya. Extreme events such as high temperatures have been linked to increased mortality, especially of older people and the infirm, in major cities.[71]

Important developments and insights in this general area of research include the following:

- The consequences of environmental change depend as much on the social systems that produce vulnerability as on the biophysical systems that cause environmental change.
- The consequences of environmental change are strongly dependent on the ability of people and social systems to adapt; consequently, access to economic resources is a key mediator between environmental changes and their impacts.
- Climate models can be linked to crop models to provide early warnings of famine.
- Human health may be an important area affected by climate change.

Knowledge is not yet adequate to achieve several goals critical to anticipating the likely consequences of future environmental changes, such as:

- Developing indicators of vulnerability that are sensitive to regional and social variations.
- Projecting vulnerability estimates into the future.
- Linking mesoscale outputs of climate models to regional impacts, taking into account vulnerability and the ability of vulnerable individuals and social systems to adapt.

Impacts of Past and Present Climate Variability

The most noticeable and perhaps most serious effects of long-term climate change may not be slow changes in average temperature or precipitation but rather such extreme events as storms, droughts, heat waves, floods, and wildfires; in some climate change scenarios, epidemics may be the most serious of all the dangers. Because of the importance of such episodes to society, environmental scientists are increasingly attempting to predict changes in the frequency of extreme weather events and to identify the boundary conditions for the spread of disease. Social scientists have explored the impacts of climate variability in historical and archeological studies and in research on the human impacts of climatic natural disasters. These studies have highlighted the importance of understanding vulnerability and adaptation.

Several recent studies suggest links between drought and the collapse of civilizations in Asia and Latin America. For example, the abandonment of settlements in the Andean altiplano and Amazon River basin has been shown to correlate with paleoenvironmental evidence of severe El Niño events.[72] A large body of work examines the influence of climatic variations on European and North Atlantic history,[73] showing the effects of the Little Ice Age and cooler periods associated with volcanic eruptions and the vulnerabilities of different social systems to those changes. As scientists gain new insights into paleoclimatic variability, rapid climate change, and shifts in decadal circulation patterns, social scientists can apply archeological and historical techniques to examine the human dimensions of these events.

Insights from natural hazards research also have great relevance to understanding the consequences of climate variability.[74] In the immediate aftermath of floods, hurricanes, and fires, social cohesion increases, and buildings and services tend to be restored with much greater speed than they were originally developed. Anticipatory responses, however, are more uneven, even under normal regimes of climate variability. For example, many vulnerable households and businesses fail to insure adequately against floods, even when warned of an increased near-term likelihood of flooding, and vulnerable communities often resist planning and zoning changes that would render them less vulnerable. If global climate change increases the frequency of so-called 100-year floods and other natural disasters or brings about disasters outside the range of previous experience, weaknesses in anticipatory responses will become more costly.

Researchers have begun to think about hazard response in terms of systems, recognizing that societies have always had systems for responding to the range of climatic variations, including extreme weather events, that they normally experience. For example, seasonal migration in drought-prone areas, crop diversification, hazard insurance systems, social norms of helping, flood control, fire management policies, zoning regulations, river monitoring, and other social and technological systems can alter the frequency, severity, extent, and distribution of economic losses associated with hazards. Such social systems can make as much difference in human outcomes as the distribution of weather events themselves, largely because of the effects of these systems on vulnerability.[75]

Global environmental change challenges human hazard management systems with potential major environmental surprises resulting from the nonlinear behavior of global environmental systems. Societies may be faced with rapid climate change events, ecological collapse, or epidemics that may be different in kind, faster in rate of onset, or greater in severity than the changes for which existing hazard management systems are prepared. We do not know how societies may alter their hazard management systems to face the prospect of such environmental surprises. However, studies of environmental risk management and decision making show that changing practices in anticipation of high-consequence/low-probability events poses a major challenge: democratic societies typically have great difficulty reaching consensus on policies to manage these sorts of environmental risks and even in arriving at widely shared understanding of them.[76]

Hazards research has also shown the potential value of understanding the consequences of El Niño and other potentially predictable aspects of seasonal to interannual climate. Several studies have linked El Niño events to major disaster losses in such regions as Northeast Brazil, Australia, and Southern Africa, and insights from these studies can contribute to applications of improved predictions of climatic conditions. For example, the correlations between crop yields and droughts associated with El Niño indicate the potential value of forecasts to food security.[77]

Important insights and findings from this area of research thus include the following: the vulnerability of a society and of its hazard management systems is often more important than the magnitude of a climatic event in determining impacts on people, and past climatic variations may have been associated with large-scale abandonment of human settlements and major migrations. More information is needed about the following issues: the consequences of newly identified rapid climate changes of the past and the ability of hazard and resource management institutions to respond to surprising shifts in climate and to seasonal forecasts.

Ozone Depletion

The main reason for widespread concern about low Antarctic ozone levels, and an important stimulus for policies to eliminate ozone-depleting chemicals, has been the concern that declines in stratospheric ozone could affect human health by increasing skin cancer, causing eye problems, and stressing immune systems. Considerable research has been directed to better understanding the links between ozone depletion and human health, as well as possible impacts on ecosystems. Most of the research to date consists of epidemiological and medical studies on the effects of increased ultraviolet (UV) radiation.

Considerable progress has been made in understanding the links between UV exposure and skin cancer, and investigating the still-controversial link of UV exposure to cataracts.[78] Additional research is needed to accomplish several critical goals: to examine the effects of ozone depletion on animal populations and ecosystems of economic or other societal value, to understand behavioral and demographic aspects of UV exposure, and to link trends in industrial use of ozone-depleting chemicals to risks for human health and ecosystems.

Environmental Change and Security

Another set of studies has examined the ways in which global change may lead to conflict, mass migration, or threats to national security. This research tends to build on studies of deforestation and climate impacts that suggest that environmental change may cause competition over resources, refugee migration, or political unrest. For example, it has been suggested that environmental degradation in developing countries has resulted and may again result in violent conflicts.[79] It has also been discussed how climate change may create conflict over international water resources and may destabilize food security.[80] A new IHDP project sets out to examine the relationship between global change and security from global to local scales.

This research focus has achieved one important end: highlighting the plausibility of several types of potential impacts of global environmental change on conflict and security. More research is needed to accomplish a related goal: providing careful empirical analysis of the relative roles of environmental change and other factors affecting conflict and migration.

What Are the Potential Human Responses to Global Change?

It is difficult to separate conceptually the causes and impacts of global change from responses to it because in many cases responses to environmental change immediately modify the causes and the impacts. For example, the adaptive responses of farmers to drought have shown how hazard warning systems moderate disaster losses. Studies have examined human responses to global change at the individual, community, national, and international levels. This section reviews progress in several pertinent areas of research, including international environmental policy, local and regional institutions, decision making and risk analysis, and valuation. We also examine progress in "integrated assessment" of global environmental changes.

International Environmental Policy

Much of the work on global-level international policy and institutions has focused on the development and implementation of regulatory "regimes" established through transnational, regional, and international agreements, such as the Montreal Protocol, the Climate Change Treaty, and the Biodiversity Treaty. These studies draw on work on the theory of international regimes, bargaining, and the structural and institutional conditions supporting international cooperation.[81] They have built the beginnings of a useful body of data and identified suggestive empirical regularities, although they have not yet supported rigorous testing of hypotheses. More specifically, these studies have focused on regime formation, the modes of influence of international environmental institutions, the use of financial transfers for international environmental protection, long-term comparative national and international policy development, the implementation of international environmental agreements, the analogy between international environmental protection and local management of common-property resources, and the systems for monitoring and enforcing compliance.[82]

A significant body of research on policy instruments at the national and subnational levels is highly relevant to implementing international environmental agreements. This research has led to improved understanding of the strengths and limitations of strategies such as regulation, various classes of financial incentives and penalties, liability law, provision of information, inducements for technological development, disaster prevention and preparedness, and alterations in the structure of markets for mitigating and responding to global change.[83]

The following are some important findings and insights of this research domain:

- Several systematic difficulties exist in changing the behavior of specific targeted nation states, even when policies are backed up with financial resources.[84]
- Policy is often strongly path dependent in that early decisions may constrain or determine later ones, thus making discussion of alternative policies extremely difficult at later stages.[85]
- Transnational networks of scientists can play a strong role in early definition and framing of an issue, although they have only limited ability to motivate international agreement or to influence the interpretation of scientific knowledge by political decision makers.[86]
- Assessment of risks and response options tends to follow, rather than lead, political target setting, and the range of options tends to contract over time.[87]
- Coercive sanctions have limited effectiveness in enforcing compliance, compared with carefully designed, linked systems involving rule design, information provision, granting of capacity and legal authority to selected actors, and transparent processes of implementation distributed among multiple formal and informal institutions.[88]

Knowledge is not yet adequate to achieve the following:

- Identify specific combinations of policy instruments and monitoring strategies that will induce a broad range of actors to behave in ways that lead to achieving internationally agreed-upon goals.
- Identify specific international and national institutions that can effectively link the international, national, and local levels and make it possible to design effective and acceptable policies.

National and Local Institutions

Human responses to global environmental change are shaped by institutions, defined broadly as the norms, regulations, interpretations or understandings, and social organizations that bear on a particular activity. The past decade has seen a renaissance of research on the structure and dynamics of social institutions and a change in thinking about how these institutions shape human activity.[89] This work is being applied increasingly to problems of institutional design for managing environmental change.[90] It is developing a knowledge base on how social institutions have succeeded and failed at long-term management of environmental resources that will be useful for informing future environmental policy choices.

The fundamental issue for environmental management is that of designing incentive-compatible institutions.[91] Such institutions are capable of internalizing, within individual households, private firms, and public organizations, the costs of the negative effects of human activities, effects such as pollution, when they are not otherwise penalized by the market. These so-called externalities are a major source of environmental stress. Even if accurate measurement is developed of the costs of externalities, the question remains of how to design institutions to avoid generating these negative effects. At present we are better at dealing with emissions and other externalities after the fact than at preventive institutional design.

An important focus of recent research on national and local institutions is property rights institutions, as they bear on the management of environmental resources.[92] One advance has been to demonstrate the ambiguity of such terms as "the commons" and "common property" and hence the need to specify property rights arrangements in more detail.[93] The term "common property" has been used to signify, at one extreme, unregulated open-access situations, as in many high-seas fisheries, frontier agricultural or mining development, and air and water pollution. At the other extreme, it has referred to highly regulated and tightly circumscribed systems governing the use of land or natural resources as found in some communities, where "commons" signifies a viable institution of collective rights and responsibilities.[94] Most "common property" institutional arrangements are somewhere in between.

Research has increasingly focused on alternatives to top-down government control over commons situations, alternatives such as collective action on the part of resource users, and market-based systems of allocation. Attention has been redirected to questions about external and internal conditions under which collective action can prevent or mitigate environmental problems. This line of research has produced significant insights, showing, for example, that common-pool resources are not inevitably destroyed by human overexploitation as presumed in the model of the tragedy of the commons. Rather, the fate of these resource pools depends on the formal and informal institutions that monitor their condition and control their use.

Comparative case studies and game theory experiments show the importance of time, scale, and socioeconomic structure to the success of collective action, and they show the need for local institutions to establish good systems of monitoring, communication, and normative control of behavior.[95] They also show the importance of appropriately structured linkages between local organizations, which tend to have more detailed knowledge about resources, and higher-level organizations, which can enforce access rules and address externalities beyond the locality.[96] These findings bear on questions of institutional reform and innovation, suggesting as they do some advantages of more decentralized management regimes.

Important advances in the knowledge of institutions include the following:

- The discovery that common-pool resources have sometimes been managed sustainably by societies for periods of several centuries.
- The development of a typology of institutions for managing common-pool resources (particularly property rights institutions).

- The development of the beginnings of a body of knowledge about how particular institutional types can effectively sustain resources.
- Recognition of the importance of locally based and self-organized institutions for monitoring and controlling resource use.
- Recognition of the ways that higher-level institutions can constrain and enable local ones.

Knowledge is not yet adequate to match particular institutional types appropriately to resource types and social settings or to identify effective ways of linking institutions across levels to maintain particular resource types.

Market Mechanisms

Property rights research also has led to research on market-based policies. Recognizing that in many situations neither local-level management nor command-and-control solutions may be effective or feasible, researchers have focused on the use of exclusive tradeable property rights or privileges in environmental management.[97] In marine fisheries, long plagued with open-access problems, individual transferable quotas have become increasingly popular. Research shows some economic benefits but mixed results with regard to creating incentives for conservation.[98] Emissions trading has been shown to reduce the costs of regulation.[99] Experiments have been under way to apply this idea internationally, for example, by using international carbon emissions offsets to help developing countries finance environmental regulation.[100] Distributional effects of trading are major issues for both fisheries and emissions regulation.

Research on modifying the ways in which markets allocate environmental resources has significant insights to offer in responding to global change. Some of this work is based on estimating the full social value of an environmental resource and instituting taxes or other financial incentives to bring the resource's price into line with its social costs. Important advances in the area of market mechanisms include indications that market-based policies (e.g., transferable quotas, emissions trading) have significant cost advantages in theory and in some applications and recognition that the effects of these policies are situation dependent. Knowledge is not yet adequate to meet several critical needs for understanding how market mechanisms would work in practice, such as estimating distributional effects of particular market-based policies and matching market-based policies to resources and social and political conditions for optimal effectiveness.

Valuation of Responses to Global Change

Responses to global change require that decision makers face important questions of resource allocation: How can decisions appropriately take into account the value of resources that have no market price? How do and how should societies allocate resources between present and future generations? How do and how should societies decide equity issues within the present generation?

Many studies use forms of cost-benefit analysis to compare the costs and benefits of various responses to global change. In these studies the economic costs of impacts as well as responses are usually included. The results of cost-benefit analyses tend to be highly sensitive to assumptions about nonmarket values, such as those of ecological services like water purification and crop pollination, human health and life, and prevention of species extinctions. Considerable effort has recently been devoted to approaches for evaluating nonmarket values by various "indirect market" methods, such as measuring the cost of activities engaged in to compensate for the effects of pollution and "hedonic pricing" (e.g., estimating the amount of increased compensation offered workers for more hazardous jobs). Each of these methods has been applied to a restricted class of nonmarket values, and each has certain clear limitations.[101] Partly to overcome the limitations of indirect market approaches, economists have increasingly turned to contingent valuation methods, in which individuals are asked directly to express their willingness to pay for particular environmental improvements. This technique has found some acceptance in policy circles, but methodological and conceptual questions about the approach are still being hotly debated.[102] Some progress has also been made in estimating the nonmarket value of environmental resources for inclusion in national income accounts.[103]

Researchers have also examined the use of novel methods for integrating disparate kinds of values without converting them to monetary units. Techniques of multiattribute utility analysis allow values to be integrated in various ways to reflect users' value priorities.[104] Simulations and policy-exercise studies clarify key uncertainties, values, and interactions, and deliberative methods involve both experts and nonexperts in interpreting analyses.[105] Such experiments promise to yield methods that complement economic techniques of valuation, where the latter give incomplete or inconclusive results.

There also have been efforts to deal analytically with the obligation of the present generation to future generations. For example, a choice not to mitigate climate change benefits the current generation but imposes unknown adaptation costs on future generations. These analyses are highly sensitive to the discount rate used to represent tradeoffs between present and future resources. Critics of conventional cost-benefit analysis often insist that the discount rate approach results in a dictatorship of the present over the future. The issue of how to take into account obligations of the present toward the future and the problem of intergenerational income distribution both remain unresolved. Ethical analyses and discussions of these issues remain more unsatisfactory than economic analyses; however, scientific debate remains contentious.[106] Analytical techniques are also being developed to address equity issues in the present generation.[107]

This field of investigation is in a highly vigorous state. For its results to have satisfactory practical value, analysis must develop to the point at which it can give confidence in the validity of particular analytical techniques for estimating nonmarket values and provide a satisfactory way of analyzing equity issues, particularly those relating to intergenerational equity.

Understanding Risk, Uncertainty, and Complex Choices

Responding to the prospect of global change requires interventions in complex systems that are not fully understood. Decision makers can benefit from recent advances in understanding human judgment and decision processes regarding complex environmental choices. Over the past decade this research has increasingly clarified why scientific efforts to analyze and assess global environmental threats do not easily lead to social consensus on policy responses to those threats. This research shows that, while analyses normally focus on a few critical outcomes, such as sea level rise or species extinction, nonexperts commonly consider multiple dimensions of environmental conditions and decisions, such as risks to human health, economic costs and benefits, the extent of scientific uncertainty and ignorance, catastrophic potential, and threats to aspects of the environment with intrinsic value. Moreover, even single dimensions such as human health are multifaceted; people assess risks partly depending on whether they believe they can control personal exposure and their own emotional responses to the specific hazard.[108] The factors that matter most to people can vary with the situation and with their social position, and these differences then influence their policy preferences.[109]

These findings have implications for scientific choices and the policy process. Chief among them is the idea that the information that results from science-driven research agendas is not necessarily considered useful or relevant by those whose decisions the scientific analysis is intended to inform. For scientists to know what information will be considered useful and relevant, they must have input from those who participate in environmental decisions. Each technique used to assess environmental risks inevitably makes judgments about what the problem is that needs scientific input, which dimensions of the problem should be investigated, and their relative importance.[110] Consequently, decisions that rely on any particular analytical technique are often rejected by people who do not accept its underlying judgments. Moreover, decisions made without the participation of some of the interested and affected parties tend to be rejected by those parties. Consequently, decision processes that are too narrowly based, either in terms of analysis or participation, often fail to meet decision makers' needs for information and involvement.[111]

The ability to reach an implementable decision depends on the process that combines analysis and deliberation to frame scientific questions, gather and interpret information, and present it to participants in the decision in ways that address their needs for information and understanding. The critical elements of this decision process and some important related research questions have been identified.[112] The decision process is also a major concern of the new National Center for Environmental Decision-Making Research, established by NSF. Additionally, research has begun to illuminate how scientific and technical information is incorporated into environmental decision making at local, national, and international levels.[113]

A number of important findings have been made in this research area:

- Whereas many risk assessments consider only a few dimensions of risk (e.g., mortality risk, economic loss), nonspecialists' judgments about risk typically consider multiple dimensions.
- Objections to the results of risk assessments often arise from disagreement about judgments underlying the assessments or from restricted participation in making those judgments.
- The adequacy and acceptability of a judgment about risk depend on both the underlying analysis and the deliberative process that judged which analysis to do, how to collect information, and how to interpret it.

Knowledge is not yet adequate in this field to accomplish several essential tasks, such as:

- Adequately characterizing uncertainties and scientific disagreements about the nature and extent of risks.
- Designing processes that combine analysis and deliberation to ensure that information is developed and organized to meet the needs of the range of decision participants.
- Structuring procedures that inform scientists' work and decision makers' understanding with a combination of formal analysis and the information, perspectives, and judgments of others involved in risk decision making.

Integrated Assessment

One approach used to understand the implications of policy responses to global change is known as "integrated assessment." In integrated assessment, methods or processes are applied to combine knowledge from multiple domains, such as socioeconomic and biophysical fields, within a consistent framework to inform policy and decision making. Integrated assessments of environmental issues have been conducted since the 1970s,[114] but the past 10 years have seen a flood of interest and activity, particularly to address global climate change. Since 1990 the number of integrated assessment projects relating to climate change has grown from only a few to more than 40.[115]

Although the concept of integration has been very broadly applied with regard to what is integrated and how, recent practice in the area of climate change has been rather narrow. Integration can mean "end-to-end" connection of a causal chain from fossil fuel emissions and land use to their impacts, with weighing of climate change impacts against measures to reduce them or to adapt. (Some amount of such "vertical" integration is often taken as a requirement for integrated assessment.) Integration can also denote expanding each link of this chain to consider more diverse source activities and emissions, more atmospheric and biotic processes, more forms and sectors of impacts, or more spatial detail or heterogeneity of agents. Integrated assessments may also examine social and biophysical linkages between climate change and other issues (e.g., ozone depletion, tropospheric air pollution) or include linkages to other social or policy issues such as public health or economic development. In addition to formal modeling, methods for integration can also include structured cross-disciplinary discourse; judgmental integration of data, theory, and formal models from separate domains; and structured heuristic processes such as simulations, scenario exercises, and policy exercises.

A major purpose of integrated assessment is to provide a consistent framework for the representation, propagation, analysis, and communication of uncertainties. A striking result of the few attempts to integrate uncertainty quantitatively across biophysical and socioeconomic domains has been that, among the various kinds of uncertainties, socioeconomic uncertainties appear to predominate in assessing aggregate impacts and net benefits of policies and decisions. Key socioeconomic uncertainties include future population growth and migration, social and political determinants of environmentally relevant consumption, rate and character of technological change, adaptation-mediated regional impacts of climate and environmental change, effects of policies, and variation in preferences.

For example, in an early assessment that integrated energy-economic and carbon-cycle models, it was found that the largest contribution to uncertainty in atmospheric CO_2 concentrations at the end of the next century came from estimates of the ease of substitution of fossil and nonfossil energy inputs in the economy and general productivity growth;[116] uncertainty in the airborne CO_2 fraction and in total fossil fuel resources ranked near the bottom of all contributions to uncertainty. In 1993 and 1996, studies using a stochastic integrated-assessment model found that differences in preferences dominated climate uncertainty in determining preferred policy choice.[117]

Recent integrated climate assessment, however, has stressed vertical, or end-to-end, integration, primarily by building single integrating computer models. These models typically combine and modify preexisting models of energy and the economy, atmospheric chemistry and dynamics, oceans, the terrestrial biosphere, and/or agriculture and other forms of land use. In each project some domains are represented richly, others very schematically. Most integrated assessment projects have a national to global scale, rather coarse spatial and sectoral resolution, and weak representation of policies and political processes.

Early work on integrated assessment of climate change combined energy-economic models with either accounting or input-output systems to develop comprehensive emissions scenarios or with simple highly parameterized atmospheric models to project the effect of specific economic and control scenarios on atmospheric trace gas concentrations.[118] More recent projects have added climate and impacts modules.

Projects differ in their conceptual emphasis and the potential insights they can offer. Some concentrate on the dynamics of emissions, atmospheric accumulation, impacts, and responses. These projects postulate a single global optimizing producer-consumer and require a common metric for abatement costs and climate damages, so they normally represent regional or global climate damages by simple aggregate functions of temperature change. These models allow the investigation of dynamically optimal abatement strategies that balance, over time, the costs of emissions abatement and damages from climate change or that meet a specified environmental target at minimum cost. They also permit study of how preferred policies depend on alternative specifications of damage functions, discount rates, the dynamics of impacts and technological change, or the structure of world regions and of bargaining.[119]

Other integrated assessment projects concentrate on the specification and propagation of uncertainty, allowing identification and ranking of key policy-relevant uncertainties or the elaboration of adaptive and learning strategies for responding to progressively resolved uncertainty over time.[120] Still other projects concentrate on the elaboration of spatial and sectoral detail for climate impacts, human adaptation and responses, and human-mediated feedbacks through land use change to the climate system.[121]

Integrated assessment practitioners have claimed insights such as the following: that a large near-term abatement effort for climate change is not justified; that the market impacts of climate change in high-income countries (but not low-income ones) will be small; that optimal abatement paths would reduce gross domestic product by only a few percent, compared with unconstrained paths, and can be accomplished with carbon taxes of a few dollars per ton; and that delays of a few decades in controlling emissions are preferable to immediate action, even if stringent reductions are subsequently determined to be needed.[122] These conclusions, however, depend on several particular characteristics of most assessment models: they offer very limited representation of the possibility of extreme events; they only reference doubled CO_2 scenarios and so fail to include the concentrations likely by the end of the next century under aggressive fossil fuel growth, which drives atmospheric, ecosystem, and impacts models all far out of their validated ranges; they include weak or no representation of multiple interacting environmental stresses; and they assume limited learning in technological change or policy.

Important advances in knowledge from integrated assessment modeling include the following:

- The finding that socioeconomic uncertainties dominate biophysical uncertainties in contributing to aggregate uncertainty about future climate impacts and preferable response strategies.[123]
- Initial quantitative estimates of the benefits available from various levels of international cooperation to manage climate change.
- An evaluation of the implications of sulfate aerosols in climate change for alternative abatement strategies.
- A preliminary characterization of the effects of linked demographic, economic, and climatic pressures on land cover and atmosphere.[124]

Knowledge is not yet adequate in this field to achieve the following:

- Reduce major socioeconomic uncertainties in integrated assessment models.
- Estimate impacts and preferable policies from models that relax some of the most important restrictive assumptions of existing models (e.g., doubling of CO_2 concentrations).
- Provide acceptable techniques for choosing among model simplifications, so that outputs best meet users' needs.

What Are the Underlying Social Processes, or Driving Forces, Behind the Human Relationship to the Global Environment?

Human dimensions research has also examined fundamental questions about the broader political, social, technological, and economic forces that shape the human activities that cause environmental change and influence its consequences. The number of such forces that may directly or indirectly alter the global environment has no limit. This section focuses on several driving forces about which important scientific progress has been made and which are often mentioned as arenas for policies to mitigate environmental change. There are many other important social forces and phenomena whose direct or indirect environmental effects may also be large and that may also have policy significance. These include national taxation policies, economic inequality within and between countries, war and the international arms trade, and societies' treatment of women. Important scientific progress has been made in understanding how humans perceive global change; the ways that individuals and institutions cope with environmental changes; and the dynamics of human populations, technological change, and economic transformations.

Public Attitudes and Values

Public support is necessary for any collective response to global environmental threats, whether through policy decisions or the aggregated actions of large numbers of individuals and organizations. A series of studies shows strong and persistent concern and support for environmental quality and protection in a variety of countries, rich and poor;[125] in the United States and other countries where relevant data are available, this support cuts across socioeconomic lines. In some developed countries, concern is strongly correlated with education; in some it is strongest in younger age cohorts. Concern about global environmental problems relative to local and national ones is strongest in developed countries, whereas in countries with highly visible pollution problems, environmental issues closer to home are seen as relatively more serious.[126] Environmental concern is strongest in countries with serious objective pollution problems and in countries with strong environmentalist values.[127]

Research on the factors underlying environmental concern finds that it is partly rooted in basic psychological values, particularly concerns with the welfare of others and of future generations and a widespread belief in the sacredness of nature.[128] This work draws on extensive basic research that has developed a comprehensive typology of human values.[129] Additionally, environmental concern reflects beliefs about how environmental conditions may affect those things that an individual values, suggesting that public response to newly identified environmental conditions may depend on the kinds of consequences projected for those conditions.[130] Despite some widely held misconceptions about the causes of climate change,[131] such variation from accepted scientific accounts does not seem to diminish levels of public concern with the environmental problems that also concern scientists.

The other side of the coin of environmental concern is an apparent unconcern by individuals about the environment, as reflected in increasing levels of materials and energy consumption associated with increased income. Critics of "consumer society" point to advertising and the mass media as drivers of materialist attitudes and desires and argue that these forces and others are driving the emerging middle classes in many developing countries to emulate North American styles of consumption. These plausible arguments have not yet been supported by careful quantitative studies of the relevant social forces, attitudes, and behaviors.[132]

Important advances in knowledge in this area are documentation of widespread support for environmental protection across countries and socioeconomic groups and initial identification of the ways that values, beliefs, and attitudes affect political support for environmental policy. Knowledge is not yet adequate to relate the development of public attitudes to mass media coverage of environmental issues and the roles of elites, interest groups, advertising, and social movement organizations and to model the development of public support for action on emerging global environmental issues as a function of new scientific knowledge.

Individual and Household Behavior

Household consumption of energy and certain materials is important both in causing and in responding to global change. Consumer behavior is determined partly by values, attitudes, and beliefs but is strongly mediated by nonattitudinal factors, including the cost and inconvenience of making environmentally significant behavioral changes, the availability of relevant technologies, institutional barriers, knowledge about which behaviors are effective, and the presence or absence of supporting public policies and social pressures. Consequently, the determinants of consumption are highly situation specific, and efforts to change the environmentally relevant consumption of households require multifaceted approaches that identify and address the barriers to change that are most important for the specific behavioral change and target actor.[133] Considerable progress has been made in understanding certain key classes of consumption, such as residential energy use in some high-income countries. A major research challenge, only now beginning to be addressed, is to understand how the factors that drive such consumption vary with national and cultural context.

Political behavior is also important to responses to global change. As in the case of consumption, the connections between individual concerns and political influence are complex and imperfect. Political action reflects opportunities for effective political participation individually and through environmental organizations, changing value priorities, the framing of issues in the mass media and by interested parties, and the actions of scientific experts individually and through epistemic communities.[134] Research linking environmental attitudes to political participation and influence is helping build understanding of the political feasibility of policies to meet international commitments.

Important advances in knowledge of individual and household behavior include the following:

- Improved understanding of the many factors affecting specific types of environmentally significant consumption at the household level (especially energy use) in high-income countries and recognition of the situation specificity of these effects.
- Recognition of the various factors affecting individuals' political behavior on environmental issues.
- Appreciation of the need for multifaceted approaches in policies aimed at altering consumption patterns.

Knowledge is not yet adequate to achieve several ends:

- Project environmentally significant consumption in developing countries as a function of economic, demographic, and other changes.
- Model the causes and trajectories of environmentally significant household consumption other than energy.
- Develop more realistic assessments of likely environmental policy outcomes that take behavioral responses into account.

Economic Transformations

Various large-scale economic transformations around the world may have major implications for the generation of environmental change and for human vulnerability to it. These transformations include the dependence of an increasing proportion of the world's population on global markets for necessities such as food and fuel that were previously produced locally, much of them outside the money economy; increasing liberalization of international trade;[135] the emergence of service economies in place of manufacturing-based ones in most high-income countries; and the transformation of formerly socialist economies from a central command model to a more decentralized market-based one.

One of the most important themes in the past 10 years of social science research has been the "globalization" of economies and cultures.[136] The increasing mobility of capital and labor has facilitated the expansion of transnational corporations and massively restructured the geography of industry, agriculture, human settlements, and all of their associated environmental impacts.[137]

The environmental effects of trade liberalization are more complex than sometimes realized. Despite claims that trade liberalization has predictably negative environmental impacts, the limited existing evidence suggests that environmental impacts are sometimes positive (e.g., better allocation of soil and water resources in agriculture) and sometimes negative (e.g., foreign investment in countries with lax environmental regulations). Analyses of overall impacts must consider the effects on resource allocation, the scale of overall economic activity, the composition of output (e.g., manufacturing vs. services), effects on developing "green" technologies, and the interactions of trade with policy.[138]

The North American Free Trade Agreement (NAFTA) stimulated some important work on the environmental implications of changing trading regimes. Although some scholars claimed that NAFTA would result in improved environmental conditions, especially in Mexico, others suggested that free trade would result in environmental degradation as communities relaxed regulations to attract industry or as polluting industries moved to Mexico to take advantage of lower wage rates.[139] NAFTA was also predicted to alter agricultural production patterns in ways that would increase Mexico's vulnerability to U.S. droughts.[140]

Perhaps the most important and dramatic change in the global political economy in the past 10 years is the collapse of the Soviet bloc and the transformation of Eastern European economies. These economies had previously been among the most energy and pollution intensive in the world. Studies showed that in the immediate aftermath of political changes in such countries as Russia and Poland, greenhouse gas emissions decreased as industrial production and consumption fell in the ensuing economic crisis.[141] Now, as foreign investment and privatization transform these economies, the implications for the global environment in terms of emissions, land use, and resource management are unclear.

Important advances have been made in understanding the effects of economic transformations:

- Most of the world's food is now produced within a global system, in which most of the basic grain on the world market is produced in very few countries. The fact that many countries depend on food imports greatly enhances the regional and global impacts of climatic change and variation in those grain-producing regions on which much of the world depends.
- Industrial production is shifting from core industrial countries to the developing world.
- The service sector has grown dramatically, especially in urban areas, contributing to increased vulnerability of human settlements, as poor people move into cities for work and must often live in hazard-prone environments.

Knowledge is still inadequate for several needs:

- Establishing the theoretical and empirical links among economic globalization, global environmental change, and the consequences of global change.
- Estimating the net overall and regional environmental effects of trade liberalization.
- Estimating the likely long-term environmental effects of ongoing economic transformation in the former socialist bloc.

Human Population Dynamics

The past decade has seen substantial progress in understanding fundamental population processes: fertility, mortality, and migration as well as the relationships among them that determine population growth, age structure, and geographic distribution. This research is important to global change because population dynamics are some of the most important driving forces behind land use change and energy use and a factor in increasing demands for food, water, and living space that increase vulnerability. Efforts to reduce fertility (i.e., the number of births per woman) have received the most research and policy attention, with Asia, the Middle East, Africa, and Latin America being the areas of interest.

Based on evidence from censuses, the World Fertility Survey, the Demographic and Health Surveys, and other surveys, it is now conclusive that fertility rates have dropped in a sufficient number of formerly high-fertility countries to produce substantial reductions in the world's fertility. The world's total fertility rate has dropped to approximately 3.0 today—thus having achieved most of the reductions needed to reach replacement-level fertility. The Middle East and sub-Saharan Africa are still the regions with the highest levels of fertility, but even there evidence is emerging that fertility reductions have begun.

Considerable research has examined the causes of this fertility decline. Almost all countries that have achieved substantial fertility declines in the past 25 years have had concerted family planning programs. The effectiveness of these programs in reducing fertility levels, as opposed to other factors, such as rising education levels, has been rigorously debated.[142] Most agree that family planning programs have been one of many factors leading to fertility decline; the disagreement revolves around the size of the family planning program's effect.

With respect to mortality, most reductions in the past were attributable to declines in infant and child mortality. That trend is now shifting, and attention has been turning to questions of how long people might live. The debate on the limits has not been resolved,[143] but the research fueling the debate has helped to increase our focus on morbidity associated with increasing longevity and the need to have global change research include the effects of increased longevity.

Human migration is an issue of emerging importance for global change because of the possible environmental impacts of concentrated populations and the vulnerability of these populations to extreme events, especially when people are concentrated in coastal zones or floodplains.[144] Research progress in understanding migration has been hampered because accurate data are hard to acquire, and when they can be collected, they tend to be aggregated.

Finally, household size has been declining in a number of countries as affluence increases. For example, in the United States the proportion of all households with just one or two members increased from 46 percent in 1970 to 57 percent in 1995. Since households have effects on the environment from production and consumption that are somewhat independent of the number of household members, models should consider both population growth and growth in the number of households. Important findings in human population dynamics include the following two fundamental ones:

- Total fertility rates are declining worldwide, particularly in countries that have had concerted family planning programs.
- Human migration, particularly urbanization and movement to vulnerable environments, has been identified as a major potential influence on future environmental change.

Knowledge is not yet adequate to estimate the environmental effects of particular types of migration or to model environmental impacts as a function of household size and composition as distinct from population effects.

Technological Change

A major source of uncertainty in projecting future human contributions to global change and analyzing response costs is the rate at which improved technology will lead to the substitution of abundant natural resources for scarce ones and of reproducible capital for depletable resources. Economists and technologists have typically viewed technical change as widening the possibility of substitution among resources. This has frequently led to a bias in favor of assuming adaptation strategies for response rather than mitigation strategies. Ecologists and other biologists have typically regarded substitutability as being narrowly restricted. The argument about biodiversity is, in part, a reflection of these alternative views. The problem has not yet been modeled satisfactorily, nor has sufficient empirical research been conducted to test the alternative perspectives. However, dialogue between the two theoretical camps is increasing and signs of a conceptual synthesis are beginning to appear, in which the questions are formulated in terms of the relationship between rates of substitution and rates of resource consumption.[145]

Past research has documented some regularities in the time path of change in environmentally significant technologies, including rates of technology diffusion and secular trends toward so-called dematerialization and decarbonization; it has also documented variations around general time trends.[146] There has been a lively empirically based debate about the extent to which scarcity may induce innovations that reduce costs and find substitutes, a debate that may be heading toward synthesis.[147] Extensive studies have also been conducted of the conditions favoring adoption of technological innovations. This research is starting to make it possible for modelers of global change effects and builders of integrated assessment models to replace ad hoc coefficients of technological change with numbers based on empirical analysis and sound theory.

Important advances in this field include the following: identification of secular trends toward dematerialization and decarbonization of economies, along with variations around these trends; identification of factors influencing the rates of adoption of technological innovations; and identification of the substitution rate of inexhaustible resources for depletable ones as a key parameter for studies of sustainability.

Knowledge is not yet adequate to model the factors influencing variations in average rates of decline in national energy intensity and related indicators and variations around the average among industries and firms or to model the effects of environmental policies on rates of innovation in environmentally benign technologies.

LESSONS LEARNED

Human dimensions research has produced a large number of advances in knowledge about global change, as the previous section indicated in some detail. In general, recent research has refined earlier understanding of human-environment interactions in ways that will enable more accurate modeling and anticipation of global change and its impacts and better-informed policy responses. Many of the major advances in knowledge can be summarized in a few major categories.

Relative Importance of Socioeconomic Uncertainty

Several analyses clearly indicate that socioeconomic uncertainties dominate biophysical uncertainties in assessing future climate impacts. Because of the great significance to the future of environmental change of such phenomena as rates of economic growth and adoption of environmental technology, which can only be forecast with great uncertainty, relatively small improvements in our understanding of these phenomena can significantly improve the ability to anticipate and respond to the effects of climate change. The same reasoning may apply to ecological changes, the rates of which are also sensitive to socioeconomic factors, such as human demands for land and water, the long-term forecasts of which are highly uncertain.

Complex Determination of Environmentally Significant Consumption

The term "consumption" has different meanings in different scientific disciplines; research on environmentally significant consumption focuses on human activities, such as clearing forests and using fossil energy, that transform or degrade biophysical resources and thus affect things that people value. Accumulating evidence indicates that all of the environmentally significant kinds of consumption are determined by multiple factors, including such driving forces as population growth, economic and technological development, cultural forces, values and beliefs, political activity, institutions, and policies and by the interactions of these forces. Research shows that the interactions are typically specific to the type of consumption, the responsible decision maker, and the geographic and sociopolitical context. Studies are illuminating the operations of the driving forces as they affect particular types of consumption (energy, forest clearing) in particular contexts.

Human dimensions research is also advancing our understanding of some of the driving forces. It has documented the widespread global support for environmental protection and some of the reasons for that support, the many factors influencing environmentally significant consumption by individuals and households, the trend toward globalization of trade in environmentally significant goods and services, the increasing importance of human population migrations as an environmental threat, and the environmental significance of major technological trends affecting the rates of substitution of inexhaustible resources for depletable ones.

Importance of Vulnerability Analysis in Impact Assessment

Research on the impacts of past climatic variability on societies and economies shows that these impacts depend as much on the social systems that produce vulnerability as on the biophysical systems that cause environmental change. Vulnerability depends on a number of factors, including intensity of land and water use and population immigration in marginal areas, access to economic resources, infrastructure for hazard response, the health status of potentially affected populations, and the structure of the hazard management systems a society has in place to prepare for and manage environmental events. Vulnerability analyses are essential for estimating the human impacts of environmental change and variation. For instance, climate models can be linked to crop models and estimates of vulnerability to provide early warnings of famine, and ecological models can be linked to vulnerability analyses to estimate the effects of global change and climate variability on human health.

Importance of Institutional Design to Environmental Resource Management

Research on human use of common-pool resources has shown that the "tragedy of the commons" scenario is not inevitable. Tragedy has often been prevented and resources sustained over periods of generations to centuries by the design of institutions that monitor the conditions of resources locally, effectively govern access to them, establish norms of resource use and sanctions for overexploitation, and appropriately link local institutions with those at higher levels. A key to implementing effective responses to global change is to design incentive-compatible institutions, that is, institutions capable of internalizing the overall costs of environmental degradation for the individuals, private firms, and public organizations whose actions create environmental stress. Ongoing research on existing institutions and on the theory of institutional design is clarifying the conditions for successful long-term environmental resource management and the institutional structures that have been successful with particular types of resources. Better understanding of institutions that shape human interactions with the environment, these institutions' functioning, and their linkages is essential to forecasting global change and developing policy responses that reduce vulnerability as well as to effective long-term resource management.

Importance of Both Analysis and Deliberative Procedure in Environmental Decision Making

Wise environmental policy making requires good analyses of the various kinds of values and costs associated with environmental change and of the values and costs associated with available policy options. Human dimensions researchers are developing and refining analytical procedures for environmental accounting and valuation, cost-benefit analysis, and other tools to estimate the costs of global change and of policy response options. However, research also shows the dependence of these analyses on highly contestable judgments in areas where knowledge is incomplete and where value disagreements are significant, such as in estimating nonmarket values and intergenerational equity. One lesson is that, in addition to reliable analytical procedures, wise decision making depends on developing appropriate and acceptable processes for deliberating about analytical assumptions and identifying information needs among scientists, policy makers, and others interested in global change decisions. Research on the effectiveness of various deliberative procedures is in its infancy compared with research on analytical techniques.

Importance of a Broad-Based Infrastructure

A broad national and international infrastructure is developing for research and policy development on the human dimensions of global environmental change. Within the USGCRP the effort has broadened from a program of investigator-initiated studies funded by NSF to encompass some larger and more focused NSF initiatives, such as a set of human dimensions centers and teams and efforts on policy research and global change and on methods and models for integrated assessment; significant research activities by the U.S. Department of Energy (DOE) and its laboratories, the U.S. Environmental Protection Agency (EPA), the National Oceanic and Atmospheric Administration (NOAA), and the National Institutes of Health; smaller research programs at NASA, the U.S. Department of Agriculture, the U.S. Department of the Interior (DOI), and the National Institute of Child Health and Human Development; and highly promising interagency collaborations between NSF and EPA on watersheds and on valuation and decision making. It has proven especially important to find ways to meld a strong social scientific base with substantive focus on global change questions. Useful strategies have included interdisciplinary review of proposals; development of problem-focused initiatives at NSF; incorporation of strong social science input into the peer review process in mission agencies, sometimes drawing on NSF for support; and formal interagency collaborations.

It is also important to encourage international collaborations, to exchange ideas, improve access to data, engage in international comparative research, and take advantage of synergism among research efforts. The IHDP, under the auspices of the International Council of Scientific Unions and the International Social Science Council, has brought the international research community together in two major international conferences[148] and now has active core projects on land use and land cover change, industrial transformation, environmental security, and institutions. The IHDP currently provides a framework for collaboration among social scientists and coordination of national human dimensions programs. The IPCC also provides an important forum for the interchange of ideas concerning the human causes, impacts, and responses to climate change. Recently, regional networks and organizations, such as the Asia Pacific Network, the Inter American Institute, the European Community, and the System for Analysis, Research, and Training (START) are developing human dimensions research programs.

Significance of Improved Observational Methods and Data Systems

Observation, that is, the collection of data, relies on sources ranging from remote sensing platforms on satellites to social surveys. The quality of social data that serve global change research has been improved by applying cognitive laboratory techniques to the way survey questions are asked and using multilevel models and datasets to incorporate community, household, and individual factors into the same analyses. Longitudinal datasets collected by the U.S. government, such as the Residential Energy Consumption Survey, the Nationwide Personal Transportation Survey, the Consumer Expenditure Survey, the National Longitudinal Survey of Youth, the National Survey of Families and Households, and others, have permitted the use of more complex statistical models to understand underlying causal processes. These techniques have been further developed through privately conducted government-funded surveys, such as the Panel Study of Income Dynamics and the General Social Survey. New multicountry survey studies of environmental beliefs, attitudes, and consumer behavior will benefit from these advances.

Systems for linking datasets and increasing their availability provide opportunities for major advances in human dimensions research. The past 10 years have seen the establishment or linkage of several international databases of use in studying the human dimensions of global change. For example, a 1996 report[149] on the global environment provides an accessible compilation of international environmental trends at the country level from disparate sources, including various United Nations agencies and the World Bank. It is becoming the norm in the human dimensions community that data collected with federal funds should be placed in the public domain. The key issue now is implementation. The NASA-supported Social and Economic Data and Applications Center (SEDAC) is charged with making social, economic, and Earth science data available to the entire research and policy community. SEDAC also hosts a World Data Center—A, covering human interactions with the environment, and a data system for

IHDP. Environmental data must be integrated with social datasets already archived by organizations such as the Institute for Cooperative Programs in Survey Research or by researchers who maintain their own distributional mechanisms. Improved access to data via the World Wide Web, along with advances in software and metadata standards, have greatly improved the ability of researchers to search for specific types of data and then manipulate or download them.

RESEARCH IMPERATIVES

Although considerable progress has been made in understanding the human dimensions of global environmental change, there are still many unresolved questions and several important new areas for research. In the committee's review of progress, we identified many areas where knowledge was lacking or research results were inadequate. In this section we attempt to crystallize a research agenda of high-priority questions that might yield valuable information in the next 5 to 10 years, given sufficient attention and resources. These research imperatives have emerged from recent meetings and reports of National Research Council (NRC) committees with an interest in human dimensions research,[150] from a review of other national and international efforts to identify research priorities, and from consideration of the significant intellectual gaps and opportunities identified in the review above.

Some of these research imperatives directly support particular themes of the USGCRP, such as atmospheric chemistry and seasonal to interannual climate prediction (see Table 7.4). Some focus on particular elements of the traditional

TABLE 7.4 Human Dimensions Research Imperatives in Relation to Key USGCRP Science Themes

Human Dimensions Research Imperative	USGCRP Science Themes			
	Atmospheric Chemistry	Dec-Cen Climate Change	Seasonal to Interannual Climate	Ecosystems
1. Consumption	XX	XX		X
2. Technological change	XX	XX	X	X
3. Climate assessment		XX	XX	X
4. Surprises	X	XX	XX	X
5. Institutions	X	XX	X	XX
6. Land use/migration	X	X	X	XX
7. Decisionmaking/ valuation	X	XX	X	X
8. Scientific integration	XX	XX	XX	XX
9. Data links	X	X	X	X

NOTES: XX, strong relevance; X, some relevance.

TABLE 7.5   Human Dimensions Research Imperatives in Relation to Key Human Dimensions Themes

Human Dimensions Research Imperative	Key Human Dimensions Themes			
	Causes	Consequences	Responses	Driving Forces/ Social Process
1. Consumption	XX	X		XX
2. Technological change	XX	X	XX	XX
3. Climate assessment		XX		X
4. Surprises	XX	XX		X
5. Institutions	X	X	XX	X
6. Land use/migration	XX	X		XX
7. Decisionmaking/valuation	X	XX	XX	X
8. Scientific integration	XX	XX	XX	X
9. Data links	X	X	X	XX

NOTES: XX, strong relevance; X, some relevance.

framework of causes, consequences, and responses used in human dimensions research (see Table 7.5).   Others cut across several of these themes, develop understanding of fundamental social processes that affect human-environment interactions, or suggest broadening of the overall USGCRP agenda.

Social Determinants of Environmentally Significant Consumption

Previous research has identified changes in the use of land, energy, and materials as priority subjects in understanding the causes of global change. Although the driving forces for the use of these resources include population growth and technological change, in many regions the most important determinant of environmental impacts is the per capita consumption of energy and materials. Debates over the relative roles of "northern" consumption and "southern" population growth, and over the responsibility of different social groups within countries, have confounded international environmental negotiations and domestic policy development. A recent report[151] identifies the study of environmentally significant consumption as an important area for research, a point that is echoed by the recent joint statement on consumption of the National Academy of Sciences and the Royal Society of London, as well as by new research initiatives of the Organization for Economic Cooperation and Development and the European Community.

Consumption, that is, the human transformation of energy and materials, is environmentally significant "to the extent that it makes materials or energy less available for future use, moves a biophysical system toward a different state, or, through its effect on those systems, threatens human health, welfare, or other things people value."[152] Environmentally significant consumption and direct alterations of biological systems are the two main ways in which humanity affects the global environment.

Currently, the most environmentally significant consumption from a global perspective consists of the major activities that burn fossil fuels (e.g., travel, space heating and cooling, electricity production, industrial process heating) and activities that use chlorofluorocarbons, nitrogen, and certain other materials responsible for stratospheric ozone depletion, pollution of ecosystems, and other global environmental changes. Other activities (wood and water use, meat and fish consumption, toxic chemicals and waste disposal) are considered of greater environmental significance at local levels and by some groups. New information about biophysical processes can improve understanding of the relative environmental importance of consumption activities.

Two trends in consumption are of the greatest importance to global change. One is the rapid growth of consumption associated with the emergence of a global middle class—growing segments of populations, particularly in developing countries, that are able to afford the consumptive amenities of the developed world, such as motor vehicles, refrigeration of food and living space, and air travel. The other, potentially countervailing, trend is toward decreased consumption per unit affluence, particularly in wealthy countries, probably brought about by technological improvements, saturation of demand for some amenities, the increasing effectiveness of environmental movements, and shifts in the structure of economies. At the global level, the central question about consumption is whether the second trend can counteract the first before consumption causes unacceptable environmental changes. Below the global level, however, the question looks quite different from the vantage points of high-income and low-income countries and populations, which differ greatly in how much benefit they perceive from further increases in consumption.

The social determinants of environmentally significant consumption[153] include changes in human populations, development and diffusion of consumptive technologies and behavior patterns, economic resources available to households and firms, prices of fuels and equipment, human values and preferences, availability and use of information, structural change in economies, and public policies. One of the overall challenges is to understand the links between individual demand for goods and services and the ultimate environmental impacts of meeting those demands. Particular human needs and wants may be satisfied by a variety of products and processes that cause different types and magnitudes of environmental change. Understanding the connections can help in finding ways to decrease the environmental impact of meeting human needs.

Over the next decade, research on environmentally significant consumption should address major questions such as the following:

- What are the constituents and determinants of energy use and other environmentally significant consumption in countries and populations at different levels of economic development?
- How is consumption likely to change with increasing affluence in low-income countries and populations and does this change always follow the path that high-income countries and populations have followed?
- What social forces drive the most environmentally significant consumption types, such as travel, the diffusion of electrical appliances, agricultural intensification, water use, and purchases of high-energy-consuming vehicles?
- What are the relative roles of various determinants of consumption in different countries?
- What policies at the national level lead to greater attention in communities to such issues as urban sprawl, reducing the cost of home-to-work commuting, expansion of green spaces, and enhanced recycling of materials?
- Which materials transformations have the greatest environmental significance and what determines related kinds of consumption?
- What interventions can effectively alter the course of the most environmentally significant kinds of consumption?
- What determines public support for effective consumption policies and how do these factors vary across countries?

This research will involve analysis of disaggregate data on particular consumption types in relation to prices, policies, and physical infrastructure within countries; surveys of consumption behavior and related values and beliefs in households and firms; data comparisons across consumption types and countries; and experiments with interventions. It will also attempt to understand how culture, fashion, advertising, and various kinds of opportunities and constraints influence consumption and will investigate the ways in which economic and cultural globalization and corporate and government decisions increase, limit, or expand individual consumer choices.

Over the next 5 to 10 years this research can meet several critical goals:

- Improve understanding of the constituents, determinants, and time paths of energy use in developing countries, thus improving the ability to model and anticipate anthropogenic carbon emissions, use of biomass fuels, and emissions of local and regional air pollutants.
- Improve projections of emissions and pollution in high-income countries, along with understanding of the key behaviors driving those emissions.
- Improve understanding of how changes in water, food, and wood consumption influence land use and vulnerability to climate variability and change.

- Promote more realistic analyses of the policy options for achieving national and international targets (e.g., limitation of greenhouse gas emissions, control of regional air quality, more efficient use of water resources), taking into account knowledge about public acceptability and the requirements of successful implementation.

This research priority will develop basic understanding of consumption and provide insight about its causes, dynamics, and trajectories that will be essential for making informed decisions. It will also provide important support for scientific research in atmospheric chemistry and decadal to centennial climate change, in understanding how water demands create vulnerability to climate change and variation and how ecosystem pollution and food and materials consumption are drivers of land use change.

Sources and Processes of Technological Change

The rate of technological change is one of the most significant sources of uncertainty in climate models as well as in understanding future uses of land, ecosystems, and water. Moreover, development and adoption of new technologies together constitute one of the most important methods available for achieving national and international commitments to environmental protection and sustainability.

Improving emissions scenarios for greenhouse gases and other globally significant pollutants such as SO_2 and hydrogenated chlorofluorocarbons requires not only more accurate demographic trajectories but also sectoral studies of changes in consumption and technology, particularly in developing countries. The implementation and enforcement of international treaties to control ozone depletion and greenhouse gas emissions and the projection of future changes in atmospheric chemistry all will require a much more realistic and geographically disaggregated assessment of land use, technology substitution, consumer preferences, incentives, and trade than has been undertaken to date. This effort requires research ranging from studies of the industrial ecology of individual sectors, firms, and farms to analyses of corporate strategy, trading relations, and responses to regulations.

Over the next decade, research on technological change and the environment should address such questions as these:

- What factors determine variations among sectors and actors and change over time in the approximately 1 percent per year decrease in national energy intensity generally attributed to technological change?
- What factors determine average rates and variations around the average in the adoption of new production technologies that reduce inputs of energy and virgin materials per unit output?

- What factors determine the rate at which production costs of an environmentally benign technology decline as output increases?
- What have been the effects of prescriptive standards, best-available-technology rules, public recognition, and awards to encourage voluntary technology adoption and other technology-related environmental policy instruments on actual rates of innovation?

This research on technological change will involve econometric modeling of change in energy and materials productivity and case studies of technology adoption in firms and industries, changes in production processes, and responses to regulation and incentives, and case comparisons. Over the next 5 to 10 years this research can achieve the following:

- Account for a significant proportion of variance in "autonomous" energy efficiency improvements, thus enabling more accurate modeling of improvements in energy intensity and identifying factors responsible for rapid efficiency improvements in some sectors and by certain actors.
- Identify characteristics of technology-related policy instruments associated with more rapid innovation by industries.
- Identify important sources of variation in technology adoption across firms within industries.
- Identify some of the causes of "learning" in production processes.

Research on technological change, particularly research documenting rapid adoption of environmentally beneficial technology, will suggest effective policy options for encouraging beneficial technological change. It is also critical to improving the modeling and anticipation of the climatic impacts of greenhouse gases, to understanding regional changes in atmospheric chemistry, and to examining the role of technology in human impacts on terrestrial and marine ecosystems.

Regionally Relevant Climate Change Assessments and Seasonal to Interannual Climate Predictions

One of the great challenges of global change research is to make scientific information, such as the results of climate modeling and analysis and studies of vulnerability and adaptation posibilities, more relevant to decision making at the local level. Regional assessments have been identified as a priority by the IPCC, the USGCRP, and the International Research Institute for Climate Prediction. Regional assessments can be developed for scenarios of global warming, decadal climate shifts, and seasonal forecasts and have the potential to address many issues of concern to local resource managers, corporations, and citizens. This research priority is also highly relevant to scientific efforts to provide more useful seasonal to interannual climate predictions and to understand decadal to centennial shifts and changes in climate.

An important intellectual shift in climate impact assessment in recent years has been an increased focus on understanding vulnerability and adaptation. Understanding how global warming or ENSO will affect a local region is as much a question of understanding the social and economic characteristics of the region as it is of obtaining the appropriate results from a climate model. For example, drought impacts on crop production are mediated by access to adaptive technologies such as irrigation, fertilizer, and seeds, as well as crop prices and subsidies and coping mechanisms such as environmental information and insurance. Adaptive technologies and coping mechanisms vary considerably by region, sector, and social group. Thus, improved regional assessments require detailed studies of how vulnerability develops and can be reduced. Future models and economic analyses are likely to include vulnerability indicators and findings about processes that affect vulnerability.

Research priorities for understanding the impacts of long-term climate change should address some of the agendas established by the IPCC. These would include studies that take advantage of mesoscale model outputs and downscaling techniques to improve regional projections of climate impacts and detailed analysis and longer-term projections of changing vulnerability and adaptive strategies. As scientific understanding of other decadal shifts in atmospheric and ocean circulation improves, the research might also focus on climate interactions with the management of resources such as ocean fisheries, forests, and water and agricultural systems. For example, the dynamics of fisheries in the context of decadal climate shifts cannot be understood without an understanding of human pressures on marine resources.

In the area of seasonal to interannual climate prediction, an NRC panel has developed a research agenda to increase the social usefulness of such predictions.[154] As described in Chapter 5, there have been significant improvements in our ability to forecast climate 3 to 12 months in advance, especially in relation to changes in sea surface temperature and atmospheric circulation associated with ENSO. Because ENSO appears to be correlated with large impacts on agriculture, health, water resources, and ecosystems, this improved forecast capability has significant implications for people, especially when combined with information on vulnerability and adaptive responses.

The NRC panel considered such issues as measuring and monitoring the social impacts of climate variability, analyzing changes in vulnerability to climatic extremes and variations in vulnerability across social groups, and identifying opportunities and barriers for the beneficial use of seasonal forecasts, including improved understanding of interactions with markets and improved communication of uncertainties in the policy process and to forecast users. Research to make climate predictions more socially useful must be undertaken with close links to researchers and policy makers in affected regions and with frequent communication with those producing the predictions. As understanding and predictability extend to new regions, lead times, and levels of certainty, human dimensions research can provide important insights into local vulnerability and policy contexts as well as into human needs and responses to improved climate information.

Over the next decade, research to make climate predictions more useful at a regional level should focus on a number of questions, including the following:

- What are the sectoral impacts at the regional level of climate change and seasonal to interannual variations?
- Are there impact and vulnerability indicators that can be useful to detect the extent and severity of the impacts of global change on human populations?
- Can historical data be used to project future human vulnerabilities to climatic variation and change?
- How does climate change interact with other social and ecological changes to influence crop yields, water use, and other impacts?
- Can the mesoscale outputs of climate models be better linked to models predicting the regional impacts of climate change?
- How are the impacts of climate change and variability affected by the coping techniques available to vulnerable groups?
- When science can provide early warnings of possible catastrophes, how can this information be transformed effectively into public understanding and appropriate policy responses?

This research will include case studies of responses to past climate variations; quantitative analyses of the social and economic consequences of such variations, including adaptations and the distribution of impacts across regions and social groups; development and testing of vulnerability indicators against past data; building of models that project future vulnerability; development and linking of models; and analysis of responses to climate forecast information.

Over the next 5 to 10 years this research should be able to develop the following:

- Methods for linking mesoscale and other climate model outputs to models of regional water resources, agricultural production, energy needs, and health conditions.
- Assessments of the vulnerability of many regions to climate change and variation.
- Estimates of the potential regional impacts of future climate change and variability and the value of improved climate information, including seasonal forecasts.
- Improved methods for delivering climate forecast information.

Social and Environmental Surprises

Natural science research has identified and is evaluating several kinds of rapid and discontinuous environmental changes that might overwhelm human adaptive capacities, at least in some localities. They include rapid climate change events (as from a major disturbance of the North Atlantic Oscillation), major outbreaks of diseases in humans or key crop species, and rapid destruction of the reproductive capacity of key ecosystems resulting from chemical releases to the environment. The damage to society should such changes come to pass is obvious; it is less obvious how to deal best with the prospects of such changes. Societies must function and plan for the future in the face of continuing revelations from environmental science and high uncertainty about potential catastrophes. We must also deal with meta-uncertainty—not knowing how uncertain we are.

It is not just rapid environmental changes that may produce global change surprises. Social systems can also change rapidly and discontinuously in ways that may greatly alter environmental systems and human vulnerability to global change.[155] History provides a number of illustrations of such changes, including the environmental impact of European political decisions to colonize (including the rapid spread of diseases and land use changes) and the impact of regional and global warfare on resource consumption and ecosystems. More recent rapid political and economic changes also have major global change implications. For example, the collapse of the Soviet political bloc altered greenhouse gas emissions and land use, restructured trade and property rights, and altered political alliances. The environmental and social implications of such rapid change should be a research priority. Another form of dramatic social change has been the rapid spread of democratization and liberal economic policies in Latin America and Africa. These too alter human-environment relationships in unforeseen ways. As economic liberalization changes the terms of trade, land use and industrial production can change quickly, with pollution patterns shifting along with industrial relocation and deregulation and vulnerability to climatic extremes changing with the restructuring of agriculture and food systems. Democratization can transform public attitudes and open up decision-making processes to popular movements, altering the policy process responsive to global change. One of the most challenging questions is to understand how these rapid social changes may interact with rapid environmental change.

Some of the key questions about surprises for the next decade are the following:

- What are the human consequences of rapid climate changes in the past and present?
- What are the global environmental change implications of rapid political and social changes in the past and present?

- How have environmental and social surprises interacted?
- Which human activities (e.g., patterns of land use and management, chemical releases) can significantly alter the potential for major environmental surprises?
- When science can provide early warnings of possible catastrophes, how can this information be transformed effectively into public understanding and appropriate policy responses?
- How can hazard management systems, including insurance strategies, subsidies, technological investment, and warning systems, be organized to increase resilience in the face of major surprises and at what cost?
- How can society deal with the possibility that citizens will become immobilized by warnings of possible, but highly improbable, environmental catastrophes?

Research on environmental surprises should include comparative studies of past environmental catastrophes and hazard management systems, simulations that superimpose plausible hazard events on existing hazard management systems, and experiments to test responses of individuals and organizations to information about possible environmental surprises.

Research in the next 5 to 10 years can achieve the following:

- Document and analyze the environmental implications of recent rapid social changes, such as post-Soviet restructuring, democratization, trade liberalization, and resource privatization.
- In collaborations between social scientists and natural scientists, significantly elaborate the human dimensions of credible but low-probability geophysical catastrophes, ecological collapses, and disease outbreaks.
- Examine the consequences of and responses to catastrophes in the historical and prehistoric records, to identify the characteristics of hazard management systems that have been associated with effective response in the past.
- Evaluate the immobilization hypothesis and, if it is a serious threat to response, suggest ways of presenting information about possible surprises that could overcome such tendencies.
- Develop some practical approaches through which those responsible for hazard management systems can consider the implications of catastrophe scenarios for those systems.

Effective Institutions for Managing Global Environmental Change

To make effective and well-informed decisions to anticipate the threat of global environmental change, society needs better understanding of how social institutions influence environmentally significant human actions. This need can be seen from the following observations: international agreements set targets without much consideration to what is feasible; governments often set resource extraction limits at unsustainable levels; national policies often appear to local resource managers to be part of the problem; techniques for estimating the full social costs of natural resource consumption rarely result in either social consensus or policy decisions; institutional change often has unforeseen or unfair distributional impacts; and even when there is widespread agreement about a global change phenomenon among specialists, many people perceive a high level of scientific disagreement. Such difficulties afflict resource management institutions at levels from local to international.

Global and national institutions, which are at the same scale as the problems, must be better coordinated with local institutions, which are often at the same scale as the solutions. Decision makers need more information about how to achieve this coordination. They also need to develop institutional approaches for allocating environmental resources when market prices give incomplete or misleading information. Research on environmental management institutions advances our understanding of the causes, consequences, and responses to global change and should thus be given a high priority.

Over the next decade, research to meet these needs should address such questions as the following:

- What are the characteristics of effective institutions for managing global environmental change?
- What are the correlates of effectiveness for the management of international environmental and natural resources by international regimes and institutions? In particular, what are the conditions favoring effective implementation of commitments to protect biodiversity, forests, oceans, and stratospheric ozone and to prevent climate change?
- What are the implications, applicability, and limits of particular policy instruments, including market-based instruments and alterations in property rights institutions at international, national, and local levels?
- How do declaratory targets, consensus policies, and review processes interact to influence behavior and restructure the power relationships of states and nonstate actors?
- Which characteristics of national institutions are most conducive to sustainable resource management by local institutions?
- How can knowledge about the conditions for successful local resource management be applied to problems at national and international levels?

Research on these questions will demand a variety of methods, including systematic empirical study of existing regimes and institutions for managing global change issues at levels from the international to the local; conceptual studies of proposed institutional policy instruments, focusing on bargaining problems and links among international, national, and local levels; theoretical studies· of the bargaining, contracting, and principal-agent aspects of implementing commitments at higher levels by delegating substantial authority to lower-level agents (e.g., tradable permit systems, joint implementation, federal systems); institutional and political study of the applicability of institutions at lower levels of organization to the design of national and international policy instruments; quasi-experimental studies, case comparisons, and simulation studies of the effects of major changes in institutions and rules, based in part on data from archival records and the recollections of participants; and small-scale simulations and experiments.

Over the next 5 to 10 years, research on these issues can be expected to lead to a number of achievements:

- Identification of conditions, potential contributions, and pitfalls associated with specific policy instruments, such as tradable permits, and with specific designs of environmental institutions.
- Development of a larger and more consistent body of data on international institutions and regimes and on regional and local property rights and other institutions with which to conduct comparative studies of their formation, evolution, and influence.
- Identification of conditions under which particular national policies assist or impede the efforts of local resource management institutions to sustain their resources and identification of insights from the experience of local resource management institutions that are transferable or adaptable to national and international institutions.
- Identification of the contributions of process-based international review mechanisms to changed behavior.

Changes in Land Use/Land Cover and Patterns of Migration

Considerable progress is already being made in understanding land use/land cover change and changes in human population processes. All land use is local, but the forces influencing the dynamics of land use and land cover come not only from individuals, households, and communities but also from processes at regional, national, and global levels. To understand land use and land cover change requires knowledge of how forces within and beyond the individual actor combine to affect decisions, particularly the conditions conducive to land use decisions that are either destructive or restorative to the environment. We do not yet fully understand how individual perceptions, attitudes, and socioeconomic situations affect land use choices or precisely how various external conditions, such as trade and international political economy, in addition to local rules for access to resources, insurance regulations, distance to markets, infrastructure development, and other factors, interact in the calculus that people use in making decisions about resource use.

One of the current challenges in understanding land use change, as well as changes in the use of water, marine ecosystems, and other resources, is characterizing the role of population in environmental change and degradation. As noted, considerable progress has been made in understanding population-environment relationships and in explaining more basic population dynamics. There is some agreement that, in the future, migration, rather than changes in human fertility and mortality, will be the key demographic link between the two dynamic processes of land use and land cover change. Causation and feedback will probably move in both directions: environmental changes will likely cause migration, and migration will likely change the environment. Careful research into the relationships between population mobility and environmental change is also needed because of the growing popular concern with environmental refugees, the environmental impacts of immigration, and the role of population in environmental conflict and security.[156] There is very little empirical documentation of the relationships between migration and environment.

Population migrations in the United States, however, illustrate the process. There were large migrations into the Midwest from the middle of the nineteenth century until about 1920, after which time the population of the country became increasingly urban—until the past decade, that is, when, across the country, a growing number of households have moved outside of cities and either commute to work or work at home part of the time. This shift, if it continues, may have significant environmental consequences in terms of the consumption of fossil fuel and other resources and for land use and land cover.

To understand the interaction of migration patterns and land use/land cover change requires improved data and data analysis both on prior migrations and intended future migrations. Data are needed for individual and household levels, as well as for more aggregated levels. Data on migration and other social variables must be linked with biophysical data from remote and land-based sources on soils, climate, and other biophysical factors. The data must be collected and coded in such a manner that they can be geo-linked at spatial and temporal scales with resolutions appropriate for the theoretical issues addressed. The necessary temporal depth can be achieved through prospective and retrospective techniques. Retrospective approaches allow temporal depth in a cross-sectional survey; prospective approaches permit the inclusion of intentions and attitudes, which cannot be obtained retrospectively. The social science community is now in a position to collect and analyze the requisite migration data. In the past 15 years, considerable progress has been made in improving the quality of retrospective migration data by embedding its collection in a broader life history approach. More recently, advances have been made in collecting prospective migration data by incorporating the insights of social network analyses into the data collection process. Methodological aspects of geo-links to biophysical data are currently being worked out.

The potential now exists to significantly improve understanding of the inter-relationships between human spatial movements and land use/land cover change. Assuming that sufficient migration data are collected to cover both points of origin and receiving areas, we can move beyond such simple statements as "large migrations of individuals and families into a given area affect the land use/land cover patterns in that area." Such research accomplishments will involve disag-gregating both migration streams and land use/land cover change patterns so that specific attributes of migrants can be related to specific aspects of land cover and land use. For example, does the migration of young adults or the elderly have a larger impact on patterns of forest regrowth in rural areas or on water use?

Research priorities for land use studies have been established internationally through the IHDP/IGBP core project on land use/land cover change.[157] The U.S. research community should maintain collaborative ties with IHDP/IGBP, and the USGCRP should work to ensure that this collaboration is maintained. Research over the next decade should address such questions as the following:

- What are the links among land use change, migration, political and eco-nomic changes, cultural factors, and household decision making?
- What are the interrelations between migration and environmental change?
- What comparative case studies of land use and land cover change are useful for understanding and modeling land use change at regional and global scales?

This research will include efforts to map land use and land cover and will document changes over time, develop and validate classifications of land use and land cover, develop algorithms for making the classifications accurately from remotely sensed data, undertake comparative and statistical analyses of past rela-tionships between changes in social driving forces and land use and between land use and land cover, and develop and test regional and global models of land use and land cover change. Research progress will depend on remotely sensed data, which can provide key information on land cover and are needed at appropriate spatial and temporal resolutions.

To obtain data on past population migrations and other social driving forces, continued and improved access to earlier generations of remotely sensed data is imperative. Data collected for military and/or intelligence purposes need to be increasingly declassified and made available to the research community. This need applies to data sources of both the United States and the former Soviet Union. For future remote sensing instruments, fine-grain spatial resolution is critical, as is the ability to determine the height of buildings in urban areas. This research should take advantage of the development of enhanced, multilevel, multiscale, and comparative methods in the study of human communities across the planet;[158] it can also make use of Earth-observing satellites that offer 1- to 3-m resolution and that facilitate observation, archiving, and analysis of human impacts at that scale. Success will depend on collaboration between social scientists and physical scientists in developing better algorithms for analyzing the large datasets provided by fine-resolution satellites to address behavioral questions.[159] Use of remotely sensed data at this fine scale will require attention to confidentiality in archiving and can benefit from past experience with social data.[160]

Over the next 5 to 10 years research on land use issues can be expected to meet a number of goals:

- Development of datasets and comparative empirical studies on the social causes and consequences of land use and land cover change in different regions that will permit improved understanding of the relative roles of population dynamics, economics, and other factors in driving environmental change.
- An improved capability to include detailed land use and land cover information in regional- and global-scale models and the development of prototype land use models that can be validated and used to identify gaps in knowledge.
- Use of a wider range of satellite data to study human-environment interactions.
- Improved understanding of the relationship of population mobility to land use change, including the dynamics and environmental impacts of migration.

Methods for Improving Decision Making About Global Change

The link from science to policy is a major weakness in human response to global change. Although science-based understanding is essential for making informed decisions, it is not always obvious to scientists which information would be considered useful and relevant by participants in environmental decisions. It is also difficult for international, national, and local decision makers to make sense of available scientific information on complex environmental systems, much of which is uncertain or disputed and all of which is subject to change. Well-informed choices are even harder to make because they must be acceptable to decision participants who do not share common understandings, interests, concerns, or values. Research should pursue three related aims: improving methods for valuing nonmarket goods; improving analytical methods for integrating multiple types of decision-relevant information (e.g., integrated assessment models, cost-benefit analyses); and developing decision processes that effectively combine analytical, deliberative, and participatory approaches to understanding environmental choices and thus guide scientists toward generating decision-relevant information.

Over the next decade this research field should address a number of questions:

- Are there ways to improve economic assessments of the costs, benefits, and distributional effects of forecasted climate changes and variations, taking adaptive capacity into account?
- What are the best ways of communicating uncertainty, providing early warnings of food and health problems, and introducing climate information in the policy process?
- How can environmental quality be incorporated into national accounting systems, so that it can more easily be considered in the policy-making process?
- How can information about the nonmarket values of environmental resources be incorporated effectively into decision making about resource use?
- How can we better represent, propagate, analyze, and describe uncertainties and surprises in integrated assessment (e.g., integrating quantitatively specified uncertainty with subjective probability distributions, clarifying the relationship between uncertainty and disagreement)?
- What are the characteristics of institutional processes that ensure that scientific analyses are organized so as to meet the needs of the full range of decision-making participants for information and involvement?
- How can the knowledge and concerns of those participating in or affected by environmental decisions be used to inform scientists about how to make environmental information more decision relevant?
- How do expert advice and assessment influence policy, decision making, and collective knowledge of global change issues and how do policy makers interpret information about scientific uncertainty as they frame global change issues?
- How can decision-making procedures be structured to bring the quantitative and formal information embedded in assessment models together with scientific judgment and the judgments, values, preferences, and beliefs of elite and nonelite citizens in decision-making processes that meet the informational needs of the participants and are appropriate to the decision at hand?

Research to improve analytical techniques will use such methods as model development, with particular attention to the modeling and propagation of uncertainties through complex systems, dialogue among modelers using different methods for analyzing the same issues, experimental studies of methods for quantifying the nonmonetary values of environmental resources, and surveys to identify those values. Research to improve decision-making processes will use case studies and comparisons of existing systems that inform management decisions and will conduct experiments and simulations to test alternative processes, particularly methods that involve broadly based deliberative processes, for informing scientists about decision participants' information needs and for informing policy debates.

Over the next 5 to 10 years research on these issues can be expected to yield the following:

- Improved methods for describing uncertainty, scientific disagreement, and the potential for surprise in environmental systems (e.g., subjective probability distributions based on expert elicitation, discursive methods).
- A theoretically grounded understanding of the sources of apparent anomalies in expressed-preference methods of estimating nonmarket values of environmental goods and services.
- Clarification of the nature of conflicts over cost-benefit analyses and other techniques of integrating information in support of environmental policy decisions.
- Improved understanding of the conditions under which particular analytical approaches meet the needs of decision-making participants for information and involvement and the conditions under which these approaches need to be supplemented with other techniques.
- Improved ability to incorporate scientific information within deliberative processes that involve scientists, policy makers, and interested and affected parties in informing and making environmental policy decisions. An experimental effort should be undertaken to use dialogue among scientists, policy decision makers, and interested publics to identify promising research areas that would lead to information directly usable for policy. The current national assessment effort might be studied as an experiment in this sort of dialogue and used to identify some new and important research directions.

Improve Integration of Human Dimensions Research with USGCRP Science Themes and with Other International Research

As outlined in the USGCRP's (1997) *Our Changing Planet*, human dimensions research should be a component of each science theme as well as a cross-cutting issue. What is needed now is to organize the USGCRP so as to make this a reality. For each of the program's four major research themes, key human dimensions research activities relevant to that theme must be identified and supported. For example, atmospheric chemistry would include research on the consumption patterns and technologies that drive emissions-altering atmospheric chemistry, on the impacts of UV changes, and on the institutions and decision-making processes that result in the control of these emissions (e.g., the Montreal Protocol and its implementation). The theme area of seasonal to interannual climate prediction would include support for research on vulnerability to climate variations and the social implications of seasonal predictions. Decadal to centennial climate change research would incorporate research on consumption and land use changes that alter the global carbon cycle; the driving forces behind these changes; the vulnerability of water resources, agriculture, and fisheries to decadal shifts in ocean-atmosphere circulation; and the social and environmental effects of policies to limit greenhouse gas emissions. Terrestrial and marine ecosystem studies would encompass work on human causes, consequences, and responses to ecosystem changes, including an increased emphasis on ways in which institutions (in the broadest sense of property rights, laws, and markets) promote and prevent land use and ecosystem changes; integrative assessments of the interactions between natural variations and human exploitation of fisheries, grasslands, and forests; and studies of the human impacts of ecosystem changes resulting from multiple stresses (e.g., ecosystem changes and climatic changes). Some steps are currently being taken toward such integration (e.g., NOAA's effort to develop a research agenda on the human dimensions of seasonal to interannual climate prediction). Such efforts need to be encouraged and their research recommendations implemented.

Structuring support for human dimensions research only around themes defined by natural science is inadequate because certain human dimensions issues cut across all of the research themes and require crosscutting and independent research initiatives. These initiatives include those on valuing environmental quality, the problem of developing improved methods for environmental decision making, and some questions about the human driving forces of environmental change. The challenge of organizing research on these crosscutting topics is confounded by multiagency responsibilities for funding. The research priorities identified in this chapter cannot be addressed without focused and coordinated funding. NSF, the agency responsible for the largest share of designated human dimensions research funding within the USGCRP, is the agency with the most experience in engaging basic social, behavioral, and economic science expertise and in providing a strong peer review system for proposals. However, NSF funds primarily investigator-initiated and disciplinary, rather than problem-oriented and interdisciplinary, social science research. Many of the other agencies currently have very small budgets ($1 million to $3 million) devoted to human dimensions research, which restricts them to supporting research focused on the particular agency's responsibilities. The danger exists that certain critical research areas will be perceived as too crosscutting for funding by mission agencies and too interdisciplinary for funding by NSF.

Basic social science research on human dimensions administered by disciplinary programs at NSF in response to investigator-initiated proposals is very important. But support is also needed in the form of interdisciplinary review panels, interagency collaborations, and research driven by specific science plans and organized in centers of excellence to advance human dimensions research. There are good models provided by the now-defunct human dimensions review panel at NSF, the recent Human Dimensions Centers and Methods and Models in Integrated Assessment initiatives, and the joint NSF-EPA partnerships in environmental research. The last collaboration supports research on decision making and valuation in environmental policy and on water and watersheds. Similar partnerships could address the research priorities identified in this document on environmentally significant consumption (NSF, EPA, and DOE), land use change (NSF, NASA, DOI), and regional climate assessment (NSF, NOAA, NASA).

Recent developments in international human dimensions infrastructure and planning make it increasingly important for the USGCRP to support U.S. participation in new international research projects and networks. This participation will allow the USGCRP to leverage funds and research contributions from other countries and to strengthen scientific capability in the United States and in developing countries. The USGCRP should support both core activities and U.S. participation in those IHDP programs that are well planned and truly international, as well as other regional and international networks such as the Inter American Institute, Asia Pacific Network, START, and the International Research Institute for Climate Prediction as they organize high-quality human dimensions research. We are at a critical juncture in the development of many of these international programs and networks, and there are opportunities to participate in new and important research collaborations and to assist in defining international research agendas that should not be missed. Many countries are about to organize new human dimensions research initiatives and establish national advisory committees. There are important opportunities for regional and bilateral collaboration with institutions such as the European Commission (1995) through NAFTA, and with Japan and other countries making a renewed commitment to human dimensions research.

Improve Geographic Links to Existing Social and Health Data

With few notable exceptions, social science data have been collected without concern for research questions about the human dimensions of global environmental change. The data collection efforts have mainly been driven by other needs and paid for by agencies that are not part of the USGCRP. As a result, the overwhelming majority of the publicly available social science datasets, although potentially relevant to global change research, are not as well suited as they might be to this purpose. Several steps could be taken to improve this situation. First, when links from social science data to other observational sources are necessary, the USGCRP could make a marginal investment in the ongoing data collection to ensure that sufficient geographical location information is collected to permit linking with data from relevant observational platforms.

Assuming the requisite geographical linkage data are obtained, the issue of protecting the confidentiality of respondents arises. Below the country or perhaps provincial scale, virtually all social science data are obtained with the promise of protecting the confidentiality of individuals, households, organizations, and often communities. For a variety of ethical reasons it is essential that this confidentiality be maintained. Yet doing so makes it impossible to have geolinked public use files for the research community. Efforts need to be made to facilitate building a secure system that can link social science data to biophysical data to meet legitimate research needs while simultaneously protecting the confidentiality of respondents.

Solutions might include the establishment of physical places where researchers could go to do their analyses under appropriately supervised conditions or a system that involved appropriate legal safeguards backed up by enforceable penalties. The involved federal agencies need to establish a mechanism to study the problem and put an effective solution in place. Without effective solutions, scientific progress will be severely constrained.

Linkages between human health and ecosystem information can combine environmental monitoring with consequences and impact monitoring. For example, NSF's long-term ecological research site in New Mexico now traps rodents for hantaviruses, in collaboration with the Centers for Disease Control and Prevention. Data systems that integrate health outcomes with remote sensing/ geographic information system mapping can help researchers evaluate climate and land use impacts on food sources, predators, and habitats for rodents and other ecosystem changes with human consequences.

Linking social and biophysical data presumes stable funding for archiving and disseminating human dimensions data and sufficient financial resources to permit upgrading as the storage and dissemination technology changes. Both the Social and Economic Data and Applications Center and the Institute for Cooperative Programs in Survey Research have experienced substantial uncertainties and interannual fluctuations in their funding, which in turn has created problems for the research community. When the dissemination mechanism involves individual projects, mechanisms need to be in place to continue the archiving and dissemination of these datasets after the project is complete.

In addition, the time is right to carefully examine the extent to which existing and planned social science data serve the science needs of the global research agenda. This issue is discussed more extensively in Chapter 9, but brief reference is needed here. The bulk of social, economic, and health data used in human dimensions of global research is collected for other purposes, at scales well below the global level. To date, the principal federal agencies involved in collecting social science data in the United States and abroad (Census Bureau, Department of Labor, Department of Health and Human Services, and Agency for International Development) have not been part of the USGCRP. Furthermore, as discussed earlier in this chapter, human observations raise confidentiality issues that may not be present in other global change research areas. We recommend a careful review of the observational needs for human dimensions research, with careful attention to the ability to link to other observational systems, comparability across time and observational units, and confidentiality concerns.

CONCLUSION: KEY RESEARCH ISSUES FOR THE USGCRP

The USGCRP should address the human causes and consequences of global change and human responses to anticipated or experienced global change. Among the key research issues for the USGCRP should be the following:

1. Understanding the driving forces of environmentally significant consumption. The USGCRP should develop understanding of the ways in which various political, social, economic, and technological forces combine to result in major transformations of land, water, energy, and environmentally important materials. Such understanding will help to both anticipate future environmental changes and identify potentially useful interventions for mitigating environmental changes or easing adaptation to them.

2. Understanding sensitivity and vulnerability to environmental variations and changes. The human consequences of environmental change depend as much on the sensitivity and vulnerability of social systems and on their ability to adapt as on the environmental changes they experience. Thus, the USGCRP should develop understanding of both aspects of consequences. It should improve predictive models of specific environmental changes and variations that have human impacts and develop understanding of the causes and likely future trends of vulnerability and adaptive capacity. This research should include a focus on the characteristics of social systems that make them sensitive or vulnerable to particular environmental changes and on social changes that are likely to alter the sensitivities of particular human populations over time.

3. Understanding institutions and processes for informing environmental choices and managing environmental resources. People must respond to environmental change at all levels of social organization from the individual to the international. Decisions at all of these levels need to be informed by knowledge of biophysical and social processes and by understanding of human concerns. To enable more effective responses, the USGCRP should develop understanding of the characteristics of effective ways to integrate science and human concerns in informing decisions and of effective institutional forms by which human groups at all levels can monitor and manage the use of critical environmental resources. Experimental efforts should be undertaken to use dialogue among scientists, policy decision makers, and interested publics to guide scientists in developing information that will be considered useful and relevant by participants in environmental decisions.

NOTES

1. U.S. Department of Energy (1997), World Resources Institute (1996).
2. World Resources Institute (1996), Food and Agriculture Organization (1997).
3. World Resources Institute (1996).
4. World Resources Institute (1996), U.S. Department of Energy (1997).
5. Kates et al. (1985), National Research Council (1988).
6. Social Learning Group (1998).
7. Scott (1996), Bijlsma (1996).
8. Reilly (1996).
9. Intergovernmental Panel on Climate Change (1996b).
10. Ibid.
11. Caulfield (1985).
12. Moran (1981).
13. Hecht and Cockburn (1989).
14. Skole et al. (1994).
15. Pielke and Landsea (1998), Pielke and Pielke (1997), Greenpeace International (1994).
16. Blaikie (1994).
17. McMichael (1996).
18. Beamish (1995), Hutchings (1996), Brander (1996), Finlayson and McCay (1998).
19. National Research Council (1996a), National Research Council (1999).
20. Thomas (1954), Sauer (1963), Steward (1955), Glacken (1967), Turner et al. (1990), National Research Council (1993).
21. Turner et al. (1994, 1995), Moran et al. (1994, 1996).
22. Deforestation, Skole et al. (1994), Dale (1994); land use change, Rudel (1989), Entwistle et al. (1998).
23. Turner et al. (1995), Moran et al. (1994), National Research Council (1998).
24. Turner et al. (1995).
25. Riebsame (1990), Ojima et al. (1993), Parton et al. (1994).
26. Crosby (1972), McNeill (1992), Richardson (1992).
27. Balee (1994).
28. Tropical deforestation, Allen and Barnes (1985); living standard, Skole et al. (1994).
29. Hayami and Ruttan (1985), Boserup (1981).
30. Hardin (1968).
31. Netting (1981), Ostrom (1990).
32. Blaikie and Brookfield (1987).
33. Tucker and Richards (1983), Hecht and Cockburn (1989), Worster (1988).
34. Moran et al. (1994), Mausel et al. (1993), Brondizio et al. (1996).
35. Guyer and Lambin (1993), National Research Council (1998).
36. Turner et al. (1995).
37. Moran (1995), Skole et al. (1994).
38. Foresta (1992), Gillis and Repetto (1988).
39. Everett (1996).
40. Steele (1996), National Research Council (1996c).
41. Smith (1986), Sinclair (1987).
42. Folke and Kautsky (1996), Bailey et al. (1996), Meltzoff and LiPuma (1986).
43. Stonich et al. (1997).
44. McCay and Acheson (1987), Cordell (1989), Ostrom (1990).
45. Environmental impacts, Schipper et al. (1992), Schipper and Martinot (1993); other factors, National Research Council (1992), Dietz and Rosa (1997).
46. Schipper et al. (1992), Socolow et al. (1994), International Energy Agency (1997).

47. Shafik (1994), Grossman and Krueger (1995), Holtz-Eakin and Selden (1995), Dietz and Rosa (1997).
48. Schurr (1984), National Research Council (1992).
49. International Energy Agency (1997).
50. Wernick and Ausubel (1995), Nakicenovic (1996), Wernick (1996).
51. Energy use, Bohi (1981), Cropper and Oates (1992), Dillman et al. (1983), National Research Council (1984b); social group membership, Lutzenhiser (1993, 1997); individual habits, National Research Council (1984a), Lutzenhiser (1993), Gardner and Stern (1996).
52. National Research Council (1984a), Stern (1992), Gardner and Stern (1996).
53. Incentives, Bohi (1981), Cropper and Oates (1992), National Research Council (1984b), Stern et al. (1986); behavioral and informational factors, National Research Council (1984a), Stern (1992), Gardner and Stern (1996).
54. National Academy of Engineering (1989), Socolow et al. (1994).
55. Wernick and Ausubel (1995), National Research Council (1997).
56. Kinzig and Socolow (1994), Schlesinger (1997).
57. Consumption concept, National Research Council (1997), analysis of materials, National Academy of Engineering (1994), Wernick and Ausubel (1995), *Daedalus* (1996).
58. Schipper and Myers (1993).
59. Nordhaus and Yohe (1983).
60. Lave and Dowlatabadi (1993), Morgan and Dowlatabadi (1996).
61. Intergovernmental Panel on Climate Change (1996a).
62. Rosenzweig (1985), Parry et al. (1988).
63. Rosenzweig and Parry (1994), Easterling et al. (1993).
64. Reilly (1996), Rosenzweig and Parry (1994).
65. Nordhaus (1991).
66. Downing (1995), Liverman (1992, 1994b), Watts and Bohle (1993).
67. Easterling et al. (1993), Rosenberg et al. (1993).
68. Cohen (1996).
69. Cane et al. (1994).
70. Colwell (1996), McMichael (1996), Patz et al. (1996).
71. Kalkstein (1993, 1995), Kalkstein and Tan (1995).
72. Ortloff and Kolata (1993), Shimata et al. (1991).
73. Lamb (1995), Vickers (1997).
74. Mileti et al. (1995), Wenger (1985), Drabek (1986), Burton et al. (1993), Mileti et al. (1995).
75. Mileti et al. (1995), National Research Council (1999).
76. National Research Council (1996b).
77. Cane et al. (1994).
78. Bentham (1993).
79. E.g., Homer-Dixon (1991), Homer-Dixon et al. (1993), Homer-Dixon and Levy (1995), Myers (1993).
80. E.g., Gleick (1989), Liverman (1994a), Lonergan and Kavanagh (1991).
81. Krasner (1983), Levy et al. (1995), Young (1994a).
82. Regime formation, Young and Osherenko (1993); modes of influence, Haas et al. (1993), Young (1996); financial transfers, Keohane and Levy (1996); policy development, Social Learning Group, (1998); implementation of agreements, Jacobson and Weiss (1990); Victor et al. (1997); analogy between international and local levels, Keohane (1993), Keohane and Ostrom (1995); systems for compliance, Mitchell (1994), Chayes and Chayes (1995).
83. Nichols (1984), Shavell (1985), Tietenberg (1985), Baumol and Oates (1988), Cropper and Oates (1992), Peck (1993), Gardner and Stern (1996).
84. Keohane and Levy (1996).
85. Haas (1990), Thacher (1993), Parson (1993).

86.   Haas (1993), Litfin (1994).
87.   Social Learning Group (1998).
88.   Mitchell (1994), Chayes and Chayes (1995), Victor et al. (1997).
89.   March and Olsen (1989), North (1990), Powell and DiMaggio (1991), Scott (1995).
90.   Ostrom (1990), Bromley (1992), Baland and Platteau (1996).
91.   Groves et al. (1987), North (1990, 1994).
92.   Schlager and Ostrom (1992), Hanna and Munasinghe (1995a, b), Hanna (1996).
93.   McCay and Acheson (1987), Berkes et al. (1989), Feeny et al. (1990).
94.   Levine (1986).
95.   Axelrod (1984), McKean (1992), Ostrom et al. (1994).
96.   McCay (1995), Young (1994b), Keohane and Ostrom (1995), Renard (1991).
97.   Tietenberg (1991).
98.   Squires et al. (1995), McCay (1995), Young and McCay (1995).
99.   Tietenberg (1985, 1992), Blinder (1987).
100. Described in Hempel (1996).
101. Cropper and Oates (1992).
102. Mitchell and Carson (1989), Kahneman et al. (1993), Portney (1994), Hanemann (1994), Diamond and Hausman (1994).
103. National Research Council (1994a).
104. Keeney and Raiffa (1976), Edwards and Newman (1982), von Winterfeldt and Edwards (1986), Fischhoff et al. (1984), Gregory et al. (1993), Brody and Rosen (1994).
105. Simulations and policy exercises, Brewer (1986), Parson (1996, 1997), Toth (1994); deliberative methods, Renn et al. (1993, 1995), Sagoff (1998).
106. Viscusi and Moore (1989), Cropper et al. (1994).
107. Zeckhauser (1975), Leigh (1989), Ellis (1993).
108. E.g., Fischhoff et al. (1984), Slovic (1987), National Research Council (1989, 1996b), Krimsky and Golding (1992).
109. E.g., Vaughan (1993, 1995), Flynn et al. (1994), Stern et al. (1993), Barke et al. (1997).
110. Fischhoff et al. (1981), National Research Council (1994b, 1996b).
111. National Research Council (1996b).
112. Ibid., Renn et al. (1995), Dietz and Stern (1998).
113. E.g., Jasanoff (1986), Social Learning Group (1998), National Acid Precipitation Assessment Program (1991), Ludwig et al. (1993).
114. Grobecker et al. (1974).
115. Weyant et al. (1996), Parson and Fisher-Vanden (1997).
116. Nordhaus and Yohe (1983).
117. Lave and Dowlatabadi (1993), Morgan and Dowlatabadi (1996).
118. U.S. Environmental Protection Agency (1989), Mintzer (1987), Rotmans (1990).
119. Nordhaus (1994), Peck and Teisberg (1992), Manne et al. (1993), Wigley et al. (1996), Grubb et al. (1995), Hourcade and Chapuis (1995).
120. Models focusing on uncertainties, Morgan and Dowlatabadi (1996), Hope et al. (1993), Tol (1995), van Asselt et al. (1996), models focusing on adaptive strategies, Hammitt et al. (1992), Yohe and Wallace (1996).
121. Alcamo et al. (1994).
122. Claims about market impacts, Mendelsohn and Newman (1998), Weyant et al. (1996), Yohe et al. (1996); claims about costs of abatement paths, Manne et al. (1993), Peck (1993), Nordhaus (1994), Kolstad (1996); claims about high relative costs of immediate action, Hammitt et al. (1992), Richels and Edmonds (1995), Wigley et al. (1996), Yohe and Wallace (1996).
123. Nordhaus and Yohe (1983), Morgan and Dowlatabadi (1996), Yohe and Wallace (1996).
124. Alcamo et al. (1995).

125. Dunlap (1992), Jones and Dunlap (1992), Dunlap et al. (1993), Witherspoon et al. (1995).
126. Dunlap et al. (1993).
127. Inglehart (1995).
128. Stern and Dietz (1994), Kempton et al. (1995).
129. E.g., Rokeach (1973), Schwartz, (1992).
130. E.g., Stern et al. (1993).
131. E.g., confusion with ozone depletion and a strong perceived link to air pollution; Kempton (1991), Löfstedt (1992, 1995).
132. Consumer society debate, Goodwin et al. (1997), Crocker and Linden (1998); emulation debate, Wilk (1997).
133. Stern and Oskamp (1987), Kempton (1993), Lutzenhiser (1993), Gardner and Stern (1996).
134. Political participation, Dunlap and Mertig (1992), Brulle (1995); value priorities, Inglehart (1990, 1997), Stern and Dietz (1994), Kempton et al. (1995); framing of issues, Gamson and Modigliani (1989), Dietz et al. (1989), Mazur and Lee (1993); actions of scientific experts, Haas (1990, 1993).
135. Runge et al. (1994), Runge (1995).
136. Dicken (1992), Featherstone (1990), Sklair (1991).
137. Johnston et al. (1995).
138. Runge et al. (1994), Runge (1995).
139. Johnson and Beaulieu (1996).
140. Appendini and Liverman (1994).
141. de Bardeleben (1985), Vari and Tamas (1993).
142. E.g., Pritchett (1994).
143. See Hayflick (1994) for a summary.
144. Black (1994), El-Hinnawi (1985), Hugo (1996), Wood (1994).
145. Solow (1993), Arrow et al. (1995).
146. Technology diffusion, Grubler (1996); dematerializations and decarbonization, Herman et al. (1989), Wernick and Ausubel (1995), Nakicenovic (1996); time trends, Schurr (1984).
147. Induced innovation, Griliches (1957), Boserup (1965, 1981), Nelson and Winter (1982), Arthur (1994); synthesis, Ruttan (1997).
148. In the United States in 1995 and in Austria in 1997.
149. World Resources Institute (1996).
150. E.g., National Research Council (1992, 1994c, 1997).
151. National Research Council (1997).
152. Stern (1997).
153. National Research Council (1997).
154. National Research Council (1999).
155. Brooks (1986).
156. Kaplan (1994), Myers (1993), Ramlogan (1996); also see note 143.
157. Turner et al. (1995).
158. Turner et al. (1993), Moran (1995), National Research Council (1998).
159. E.g., Mausel et al. (1993), Moran et al. (1994).
160. E.g., experience with TIGER files from the U.S. census, National Research Council (1998).

REFERENCES AND BIBLIOGRAPHY

Alcamo, J., ed. 1990. *The RAINS Model of Acidification: Science and Strategies in Europe.* Kluwer Academic, Dordrecht.
Alcamo, J., G.J.J. Kreileman, M.S. Krol, and G. Zuidema. 1994. Modeling the global society-biosphere-climate system. Part 1: Model description and testing. *Water, Air and Soil Pollution* 76(March):1-35.

Alcamo, J., M. Krol, and R. Leemans. 1995. *Stabilizing Greenhouse Gases: Global and Regional Consequences*. National Institute of Public Health and the Environment (RIVM), Bilthoven, The Netherlands.

Allen, J.C., and D.F. Barnes. 1985. The causes of deforestation in developing countries. *Annals of the Association of American Geographers* 75(2):163-184.

Appendini,, K., and D.M. Liverman. 1994. Agricultural policy and climate change in Mexico. *Food Policy* 19(2):149-164.

Arrow, K., B. Bolin, R. Costanza, P. Dasgupta, C. Folke, C.S. Holling, B. Jansson, S. Levin, K. Maler, C. Perrings, and D. Pimentel. 1995. Economic growth, carrying capacity and the environment. *Science* 268(5210):520-521.

Arthur, W.B. 1994. *Increasing Returns and Path Dependence in the Economy*. University of Michigan Press, Ann Arbor.

Axelrod, R.M. 1984. *The Evolution of Cooperation*. Basic Books, New York.

Ayres, R.U., and L. Ayres. 1996. *Industrial Ecology: Towards Closing the Materials Cycle*. E. Elgar, Brookfield, Vt.

Bailey, C., S. Jentoft, and P. Sinclair, eds. 1996. *Aquacultural Development: Social Dimensions of an Emerging Industry*. Westview Press, Boulder, Colo.

Balee, W.L. 1994. *Footprints of The Forest: Ka'apor Ethnobotany—The Historical Ecology of Plant Utilization by an Amazonian People*. Columbia University Press, New York.

Baland, J-M., and J-P. Platteau. 1996. *Halting Degradation of Natural Resources: Is There a Role for Rural Communities?* Clarendon Press, Oxford.

Barke, R.P., H. Jenkins-Smith, and P. Slovic. 1997. Risk perceptions of men and women scientists. *Social Science Quarterly* 78(1):167-176.

Barrett, S. 1996. Building property rights for transboundary resources. Pp. 265-284 in *Rights to Nature*, S. Hanna, ed. Island Press, Washington, D.C.

Baumol, W.J., and W.E. Oates. 1988. *The Theory of Environmental Policy, 2d ed*. Cambridge University Press, Cambridge, U.K.

Beamish, R.J., ed. 1995. Climate change and northern fish populations. Special Publication 121. *Canadian Journal of Fisheries and Aquatic Sciences*.

Bentham, G. 1993. Depletion of the ozone layer: Consequences for non-infectious human diseases. *Parasitology* 106:S39-S46.

Berkes, F., D. Feeny, B.J. McCay, and J.M. Acheson. 1989. The benefit of the commons. *Nature* 340(July 13):91-93.

Bijlsma, L. 1996. Coastal zones and small islands. Pp. 289-234 in *Climate Change 1995: Impacts, Adaptation and Mitigation of Climate Change: Scientific and Technical Analyses*, R.T. Watson et al., eds. Cambridge University Press, Cambridge, U.K.

Black, R. 1994. Forced migration and environmental change: The impact of refugees on host environments. *Journal of Environmental Management* 42(3):261-278.

Blaikie, P.M. 1994. *At Risk: Natural Hazards, People's Vulnerability, and Disasters*. Routledge, London, New York.

Blaikie, P., and H.C. Brokfield. 1987. *Land Degradation and Society*. Methuen, London.

Blinder, A. 1987. *Hard Heads, Soft Hearts: Tough-Minded Economics for a Just Society*. Addison-Wesley, Reading, Mass.

Bohi, D. 1981. *Analyzing Demand Behavior: A Study of Energy Elasticities*. Johns Hopkins University Press, Baltimore, Md.

Bohle, H.G., T.E. Downing, and M.J. Watts. 1994. Climate change and social vulnerability: Toward a sociology and geography of food insecurity. *Global Environmental Change* 4(1):37-48.

Bongaarts, J. 1994. The impact of population policies: Comment. *Population and Development Review* 20(3):616-620.

Boserup, E. 1965. *The Conditions of Agricultural Growth*. Aldine, Chicago.

Boserup, E. 1981. *Population and Technological Change: A Study of Long-Term Trends.* University of Chicago Press, Chicago.

Brander, K. 1996. Effects of climate change on cod (Gadus murhua) stocks. Pp. 255-278 in *Global Warming: Implications for Freshwater and Marine Fish,* C.M. Wood and D.G. McDonald, eds. Society for Experimental Biology Seminar Series 61. Cambridge University Press, Cambridge, U.K.

Brewer, G.D. 1986. Methods for synthesis: Policy exercises. Pp. 455-473 in *Sustainable Development of the Biosphere,* W.C. Clark and R.E. Munn, eds. Cambridge University Press, Cambridge, U.K.

Brody, J.G., and R.A. Rosen. 1994. Apples and oranges: Using multi-attribute utility analysis in a collaborative process to address value conflicts in electric facility siting. *NRRI Quarterly Bulletin (National Regulatory Research Institute)* 15(4):629-643.

Bromley, D., ed. 1992. *Making the Commons Work: Theory, Practice, and Policy.* ICS Press, San Francisco.

Brondizio, E., E.F. Moran, P. Mausel, and R. Wu. 1996. Land cover in the Amazon estuary: Linking of TM with botanical and historical data. *Photogrammatric Engineering and Remote Sensing* 62:921-930.

Brooks, H. 1986. A typology of surprises in technology, institutions and development. In *Sustainable Development of the Biosphere,* W.C. Clark and R.E. Munn, eds. Cambridge University Press, New York.

Brulle, R.J. 1995. Environmental discourse and social movement organizations: A historical and rhetorical perspective on the development of U.S. environmental organizations. *Sociological Inquiry* 66:58-83.

Burton, I., R.W. Kates, and G.F. White. 1993. *The Environment as Hazard, 2d ed.* Oxford University Press, New York.

Cane, M.A., G. Eshel, and R.W. Buckland. 1994. Forecasting Zimbabwean maize yields using eastern equatorial Pacific sea surface temperatures. *Nature* 370:204-205.

Caulfield, C. 1985. The rain forests. *The New Yorker* 60(Jan. 14):41-83.

Chayes, A., and A.H. Chayes. 1995. *The New Sovereignty: Compliance with International Regulatory Agreements.* Harvard University Press, Cambridge, Mass.

Cohen, S. 1996. *Mackenzie Basin Impact Study Final Report.* Environment Canada, Ottawa, Canada.

Colwell, R.R. 1996. Global climate and infectious disease: The cholera paradigm. *Science* 274(5295):2025-2031.

Cordell, J. ed. 1989. *A Sea of Small Boats.* Cultural Survival Report 26. Cultural Survival, Inc., Cambridge, Mass.

Cowen, D.J., and J.R. Jensen. 1998. Urban change and development in southeastern United States. *In People and Pixels: Linking Remote Sensing and Social Science.* D.M. Liverman et al., eds. National Academy Press, Washington, D.C.

Crocker, D.A., and T. Linden, eds. 1998. *Ethics of Consumption: The Good Life, Justice, and Global Stewardship.* Rowman and Littlefield, Lanham, Md.

Cropper, M.L., and W.E. Oates. 1992. Environmental economics: A survey. *Journal of Economic Literature* 30:675-704.

Cropper, M.L., S.K. Aydede, and P.R. Portney. 1994. Preferences for life saving programs: How the public discounts time and age. *Journal of Risk and Uncertainty* 8:243-246.

Crosby, A.W. 1972. *The Columbian Exchange: Biological and Cultural Consequences of 1492.* Greenwood Publishing Co., Westport, Conn.

Daedalus. 1996. The liberation of the environment. 125(3):1-253.

Dale, V.H. 1994. *Effects of Land-Use Change on Atmospheric CO_2 Concentrations: South and Southeast Asia as a Case Study.* Springer-Verlag, New York.

De Bardeleben, J. 1985. *The Environment and Marxism-Leninism: The Soviet and Eastern German Experience.* Westview Press, Boulder, Colo.

Deudney, D. 1990. The case against linking environmental degradation and national security. *Journal of International Studies* 19(3):461-476.

Diamond, P.A., and J.A. Hausman. 1994. Contingent valuation: Is some number better than no number? *Journal of Economic Perspectives* 8:45-64.

Dicken, P. 1992. Europe 1992 and strategic change in the international automobile industry. Special issue: *Industrial Restructuring and Continental Trade Blocs. Environment & Planning* 24(1):11-31.

Dietz, T., and E.A. Rosa. 1997. Effects of population and affluence on CO_2 emissions. *Proceedings of the National Academy of Sciences* 94(1):175-179.

Dietz, T., and P.C. Stern. 1998. Science, values, and biodiversity. *Bioscience* 48(June):441-444.

Dietz, T., P.C. Stern, and R.W. Rycroft. 1989. Definitions of conflict and the legitimation of resources: The case of environmental risk. *Sociological Forum* 4:47-70.

Dillman, D., E.A. Rosa, and J.J. Dillman. 1983. Lifestyle and home energy conservation in the U.S. *Journal of Economic Psychology* 3:299-315.

Downing, T.E., ed. 1995. *Climate Change and World Food Security.* Springer-Verlag, Heidelberg.

Drabek, T.E. 1986. *Human Systems' Responses to Disaster.* Springer-Verlag, New York.

Dunlap, R.E. 1992. Trends in public opinion toward environmental issues: 1965-1990. Pp. 89-116 in *American Environmentalism: The U.S. Environmental Movement, 1970-1990,* R.E. Dunlap and A.G. Mertig, eds. Taylor and Francis, Washington, D.C.

Dunlap, R.E., and A.G. Mertig, eds. 1992. *American Environmentalism: The U.S. Environmental Movement, 1970-1990.* Taylor and Francis, Washington, D.C.

Dunlap, R.E., G.H. Gallup, and A.M. Gallup. 1993. Of global concern: Results of the health of the planet survey. *Environment* 35(9):6-22.

Easterling, W.E., III, P.R. Crosson, N.J. Rosenberg, M.S. McKenney, L.A. Katz, and K.M. Lemon. 1993. Agricultural impacts of and responses to climate change in the Missouri-Iowa-Nebraska-Kansas (MINK) region. *Climatic Change* 24(1-2):23-39.

Edmonds, J.A., M.A. Wise, and C. MacCracken. 1994. *Advanced Energy Technologies and Climate Change: An Analysis Using the Global Change Assessment Model (GCAM).* PNL-9798, UC-402. Pacific Northwest Laboratory, Richland, Wash.

Edwards, W., and J.R. Newman. 1982. *Multi-attribute evaluation.* Quantitative Applications in the Social Sciences Series, Paper No. 26. Sage, New York.

El-Hinnawi, E. 1985. *Environmental Refugees.* United Nations Development Program, New York.

Ellis, R.D. 1993. Quantifying distributive justice: An approach to environmental and risk-related public policy. *Policy Sciences* 26:99-103.

Entwistle, B., R.R. Rindfuss, and S.J. Walsh. 1998. Land use/land cover and population dynamics, Nang Rong, Thailand. In *People and Pixels: Linking Remote Sensing and Social Science.* D.M. Liverman et al., eds. National Academy Press, Washington, D.C.

European Commission. 1995. *Global Environmental Change and Sustainable Development in Europe,* J. Jager, A. Liberatore, and K. Grundlach, eds. DG X11. European Commission, Luxembourg.

Everett, J.T. 1996. Fisheries. Pp. 511-537 in *Climate Change 1995: Impacts, Adaptation and Mitigation of Climate Change: Scientific and Technical Analyses,* R.T. Watson et al., eds. Cambridge University Press, Cambridge, U.K.

Food and Agriculture Organization. 1997. Online database: HYPERLINK http://apps.fao.org/cgi-bin/nph-db.pl http://apps.fao.org/cgi-bin/nph-db.pl.

Featherstone, M. 1990. Perspectives on consumer culture: The sociology of consumption. *Sociology* 24(1):5-22.

Feeny, D., F. Berkes, B. McCay, and J. Acheson. 1990. The tragedy of the commons: Twenty-two years later. *Human Ecology* 18:1-19.

Finlayson, A.C., and B.J. McCay. 1998. Crossing the threshold of ecosystem resilience: The commercial extinction of northern cod. Pp. 311-338 in *Linking Social and Ecological Systems: Institutional Learning for Resilience*, C. Folke and F. Berkes, eds. Cambridge University Press, Cambridge, U.K.

Fischhoff, B., S. Lichtenstein, P. Slovic, S. Derby, and R. Keeney. 1981. *Acceptable Risk.* Cambridge University Press, New York.

Fischhoff, B., S.R. Watson, and C. Hope. 1984. Defining risk. *Policy Sciences* 17:123-139.

Flynn, J., P. Slovic, and C.K. Mertz. 1994. Gender, race, and perception of environmental health risks. *Risk Analysis* 14:1101-1108.

Folke, C., and N. Kautsky. 1996. The ecological footprint concept for sustainable seafood production. Unpublished paper presented at the Conference on Ecosystem Management for Sustainable Marine Fisheries, Monterey, Calif., Feb. 19-23.

Foresta, R.A. 1992. Amazonia and the politics of geopolitics. *The Geographical Review* 82(2):128-142.

Gamson, W.E., and A. Modigliani. 1989. Media discourse and public opinion on nuclear power: A constructionist approach. *The American Journal of Sociology* 95(1):1-37.

Gardner, G.T., and P.C. Stern. 1996. *Environmental Problems and Human Behavior.* Allyn and Bacon, Needham Heights, Mass.

Gillis, M., and R.C. Repetto. 1988. *Public Policies and the Misuse of Forest Resources.* Cambridge University Press, New York.

Glacken, C.J. 1967. *Traces on the Rhodian Shore.* University of California Press, Berkeley.

Gleick, P. 1989. Climate change and international politics. *Ambio* 18(6):333-339.

Goodwin, N.R., F. Ackerman, and D. Kiron, eds. 1997. *The Consumer Society.* Island Press, Washington D.C.

Greenpeace International. 1994. The climate time bomb. http://www.greenpeace.org/~climate.

Gregory, R., S. Lichtenstein, and P. Slovic. 1993. Valuing environmental resources: A constructive approach. *Journal of Risk and Uncertainty* 7:177-197.

Griliches, Z. 1957. Hybrid corn: An exploration in the economics of technological change. *Econometrica* 25:501-522.

Grobecker, A.J., S.C. Coroniti, and R.H. Cannon, Jr. 1974. *The Report of Findings: The Effects of Stratospheric Pollution by Aircraft.* DOT-TST-75-50. U.S. Department of Transportation, Climatic Impact Assessment Program. National Technical Information Service, Springfield, Va.

Grossman, G., and A. Krueger. 1995. Economic growth and the environment. *Quarterly Journal of Economics* 110(2):353-377.

Groves, T., R. Radner, and S. Reiter. 1987. *Information, Incentives, and Economic Mechanisms: Essays in Honor of Leonid Hurwicz.* University of Minnesota Press, Minneapolis.

Grubb, M., T. Chapuis, and M.H. Duong. 1995. The economics of changing course: Implications of adaptability and inertia for optimal climate policy. *Energy Policy* 23(April/May):417-432.

Grubler, A. 1996. Time for a change: On the patterns of diffusion of innovation. *Daedalus* 125(3):19-42.

Guyer, J.I., and E.F. Lambin. 1993. Land use in an urban hinterland: Ethnography and remote sensing in the study of African intensification. *American Anthropologist* 95(4):839.

Haas, P.M. 1990. *Saving the Mediterranean: The Politics of International Environmental Cooperation.* Columbia University Press, New York.

Haas, P.M. 1993. Protecting the Baltic and North seas. In *Institutions for the Earth: Sources of Effective International Environmental Protection*, P. Haas, R. Keohane, and M.A. Levy, eds. MIT Press, Cambridge, Mass.

Haas, P.M., R.O. Keohane, and M.A. Levy, eds. 1993. *Institutions for the Earth: Sources of Effective International Environmental Protection.* MIT Press, Cambridge, Mass.

Hammitt, J.K., R.J. Lempert, and M.E. Schlesinger. 1992. A sequential-decision strategy for abating climate change. *Nature* 357(6376):315-318.

Hanemann, W.M. 1994. Valuing the environment through contingent valuation. *Journal of Economic Perspectives* 8:19-43.

Hanna, S., ed.. 1996. *Rights to Nature.* Island Press, Washington, D.C.

Hanna, S., and M. Munasinghe, eds. 1995a. *Property Rights and the Environment: Social and Ecological Issues.* The World Bank and the Beijer Institute for Ecological Economics, Washington, D.C.

Hanna, S., and M. Munasinghe, eds. 1995b. *Property Rights in a Social and Ecological Context: Case Studies and Design Applications.* The World Bank and the Beijer Institute for Ecological Economics, Washington, D.C.

Hardin, G. 1968. The tragedy of the commons. *Science* 162:1243-1248.

Hayami, Y., and V. Ruttan. 1985. *Agricultural Development: An International Perspective.* Johns Hopkins University Press, Baltimore.

Hayflick, L. 1994. *How and Why We Age.* Ballantine Books, New York.

Hecht, S.B., and A. Cockburn. 1989. *Fate of the Forest: Developers, Destroyers and Defenders of the Amazon.* Verso, New York.

Hempel, L.C. 1996. *Environmental Governance: The Global Challenge.* Island Press, Washington, D.C.

Herman, R., S.A. Ardekani, and J.H. Ausubel. 1989. Dematerialization. Pp. 50-69 in *Technology and Environment.* National Academy Press, Washington, D.C.

Holtz-Eakin, D., and T.M. Selden. 1995. Stoking the fires? CO_2 emissions and economic growth. *Journal of Public Economics* 57:85-101.

Homer-Dixon, T.F. 1991. On the threshold: Environmental changes as causes of acute conflict. *International Security* 16(2):76.

Homer-Dixon, T.F., and M.A. Levy. 1995. Environment and security. *International Security* 20(3):189.

Homer-Dixon, T.F., J.H Boutwell, and G.W. Rathjens. 1993. Environmental change and violent conflict. *Scientific American* 268(2):38-45.

Hope, C., J. Anderson, and P. Wenman. 1993. Policy analysis of the greenhouse effect: An application of the PAGE model. *Energy Policy* 21(March):327-338.

Hourcade, J., and T. Chapuis. 1995. No-regret potentials and technical innovation: A viability approach to integrated assessment of climate change. *Energy Policy* 23(April/May):433-446.

Hugo, G. 1996. Environmental concerns and international migration. *International Migration Review* 30(1):105-131.

Hutchings, J.A. 1996. Spatial and temporal variation in the density of Northern Cod and a review of hypotheses for the stock's collapse. *Canadian Journal of Fisheries and Aquatic Sciences* 53:943-962.

Inglehart, R. 1990. *Culture Shift in Advanced Industrial Society.* Princeton University Press, Princeton, N.J.

Inglehart, R. 1995. Public support for environmental protection: Objective problems and subjective values in 43 societies. *Political Science and Politics* 15:57-71.

Inglehart, R. 1997. *Modernization and Postmodernization: Cultural, Economic and Political Change in 43 Societies.* Princeton University Press, Princeton, N.J.

Intergovernmental Panel on Climate Change. 1996a. *Climate Change 1995: Impacts, Adaptation and Mitigation of Climate Change: Scientific and Technical Analyses,* R.T. Watson et al., eds. Cambridge University Press, Cambridge, U.K.

Intergovernmental Panel on Climate Change. 1996b. *Economic and Social Dimensions of Climate Change 1995,* J. Bruce et al., eds. Cambridge University Press, Cambridge, U.K.

International Energy Agency. 1997. *Indicators of Energy Use and Efficiency: Understanding the Link Between Energy and Human Activity.* International Energy Agency, Paris.

Jacobson, H.K., and E.B. Weiss. 1990. Implementing and complying with international environmental accords: A framework for research. Unpublished paper presented at the annual meeting of the American Political Science Association, San Francisco.

Jasanoff, S. 1986. *Risk Management and Political Culture.* Russell Sage Foundation, New York.

Johnson, P., and A. Beaulieu, eds. 1996. *The Environment and NAFTA: Understanding and Implementing the New Continental Law.* Island Press, Washington, D.C.

Johnston, R.J., P. Taylor, and M. Watts. 1995. *Geographies of Global Change: Remapping the World in the Late Twentieth Century.* Blackwell, Oxford.

Jones, R.E., and R.E. Dunlap. 1992. The social bases of environmental concern: Have they changed over time? *Rural Sociology* 57:28-47.

Kahneman, D., I. Ritov, K.E. Jacowitz, and P. Grant. 1993. Stated willingness to pay for public goods: A psychological perspective. *Psychological Science* 4:310-315.

Kalkstein, L.S. 1993. Health and climate change—direct impacts in cities. *The Lancet* 342:1397-1399.

Kalkstein, L.S. 1995. Cities could face killer heat waves. *Science* 267:958.

Kalkstein, L.S., and G. Tan. 1995. Climate change and human health: International implications. Pp. 124-145 in As *Climate Changes: International Impacts and Implications,* K.M. Strzepek and J.B. Smith, eds. Cambridge University Press, New York.

Kaplan, R.D. 1994. The coming anarchy: How scarcity, crime, overpopulation, tribalism, and disease are rapidly destroying the social fabric of our planet. *The Atlantic Monthly* 273(2):44-64.

Kates, R.W., J.H. Ausubel, and M. Berberian, eds. 1985. *Climate Impact Assessment: Studies of the Interaction of Climate and Society.* Published on behalf of the Scientific Committee on Problems of the Environment of the International Council for Scientific Unions. John Wiley & Sons, New York.

Keeney, R.L., and H. Raiffa. 1976. *Decisions with Multiple Objectives: Preferences and Value Tradeoffs.* John Wiley & Sons, New York.

Kempton, W. 1991. Lay perspectives on global climate change. *Global Environmental Change: Human and Policy Dimensions* 1:183-208.

Kempton, W. 1993. Will public environmental concern lead to action on global warming? *Annual Review of Energy and Environment* 18:217-245.

Kempton, W., J.S. Boster, and J.A. Hartley. 1995. *Environmental Values in American Culture.* MIT Press, Cambridge, Mass.

Keohane, R.O. With the collaboration of E. Ostrom and M. McGinnis. 1993. Linking local and global commons: Monitoring, sanctioning, and theories of self-organization in common pool resources and international regimes. Pp. 1-15 in *Proceedings of a Conference on Linking Local and Global Commons,* R.O. Keohane, M. McGinnis, and E. Ostrom, eds. Center for International Affairs, Harvard University, Cambridge, Mass.

Keohane, R.O., and M.A. Levy, eds. 1996. *Institutions for Environmental Aid.* MIT Press, Cambridge, Mass.

Keohane, R.O., and E. Ostrom, eds. 1995. *Local Commons and Global Interdependence: Heterogeneity and Cooperation in Two Domains.* Sage, London.

Kinzig, A.P., and R.H. Socolow. 1994. Human impacts on the nitrogen cycle. *Physics Today* 47(11):24-31.

Kolstad, C.D. 1996. Learning and stock effects in environmental regulation: The case of greenhouse gas emissions. *Journal of Environmental Economics and Management* 31:1-18.

Krasner, S., ed.. 1983. *International Regimes.* Cornell University Press, Ithaca.

Krimsky, S., and D. Golding. 1992. *Social Theories of Risk.* Praeger, Westport, Conn.

Lamb, H.H. 1995. *Climate, History and the Modern World.* Routledge, New York.

Lave, L.B., and H. Dowlatabadi. 1993. Climate change: The effect of personal beliefs and scientific uncertainty. *Environmental Science and Technology* 27(10):1962-1972.

Leigh, J.P. 1989. Compensating wages for job-related death: The opposing arguments. *Journal of Economic Issues* 23:823-839.

Levine, B.L. 1986. The tragedy of the commons and the comedy of community: The commons in history. *Journal of Community Psychology* 14:81-99.

Levy, M.A., O.R. Young, and M. Zuern. 1995. The study of international regimes. *European Journal of International Relations* 1:267-330.

Litfin, K.T. 1994. *Ozone Discourses: Science and Politics in Global Environmental Cooperation.* Columbia University Press, New York.

Liverman, D.M. 1992. The regional impacts of global warming in Mexico: Uncertainty, vulnerability and response. Pp. 44-68 in *The Regions and Global Warming*, J. Schmandt and J. Clarkson, eds. Oxford University Press, Oxford, England.

Liverman, D.M. 1994a. Environment and security in Mexico. In *Mexico: In Search of Security*, B.M. Bagley and S. Aguayo Quezada, eds. Transaction Publishers, New Brunswick, N.J.

Liverman, D.M. 1994b. Vulnerability to global environmental change. Pp. 326-342 in *Environmental Risks and Hazards*, S. Cutter, ed. Prentice-Hall, Englewood Cliffs, N.J.

Löfstedt, R.E. 1992. Swedish lay perspectives on global climate change. *Energy and Environment* 3:161-175.

Löfstedt, R.E. 1995. Lay perspectives concerning global climate change in Vienna, Austria. *Energy and Environment* 4:140-154.

Lonergan, S.C., and B. Kavanagh. 1991. Climate change, water resources and security in the Middle East. *Global Environmental Change* (Sept.):272-290.

Ludwig, D., R. Hilborn, and C. Walters. 1993. Uncertainty, resource exploitation, and conservation: Lessons from history. *Science* 260(April 2):17.

Lutzenhiser, L. 1993. Social and behavioral aspects of energy use. *Annual Review of Energy and the Environment* 18:247-289.

Lutzenhiser, L. 1997. Social structure, culture, and technology: Modeling the driving forces of household energy consumption. Pp. 77-91 in *Environmentally Significant Consumption: Research Directions*, P.C. Stern et al., eds. National Academy Press, Washington, D.C.

Manne, A.S., R. Mendelsohn, and R.G. Richels. 1993. MERGE: A model for evaluating regional and global effects of GHG reduction policies. *Energy Policy* 23(1):17-34.

March, J.G., and J.P. Olsen. 1989. *Rediscovering Institutions: The Organizational Basis of Politics.* The Free Press, New York.

Mausel, P., Y. Wu., Y. Li, E. Moran, and E. Brondizio. 1993. Spectral identification of successional stages following deforestation in the Amazon. *Geocarto International* 8:11.

Mazur, A., and J. Lee. 1993. Sounding the global alarm: Environmental issues in the U.S. national news. *Social Studies of Science* 23:681-720.

McCay, B.J. 1995. Social and ecological implications of ITQs: An overview. *Ocean and Coastal Management* 28(1-3):3-22.

McCay, B.J., and J.M. Acheson. 1987. *The Question of the Commons: The Culture and Ecology of Communal Resources.* University of Arizona Press, Tucson.

McCay B.J., and S. Jentoft. 1996. Unvertrautes Gelande: Gemeineigentum Unter der Sozialwissenschaftlichen Lupe (Uncommon Ground: Perspectives on Common Property). Kölner Zeitschrift für Soziologie und Sozialpsychologie (*Cologne Journal of Sociology and Social Psychology*) 36(Fall):272-291.

McGoodwin, J.R. 1990. *Crisis in the World's Fisheries: People, Problems and Politics.* Stanford University Press, Stanford, Calif.

McKean, M.A. 1992. Success on the commons: A comparative examination of institutions for common property resource management. *Journal of Theoretical Politics* 4(3):247-281.

McMichael, A.J. 1996. Human population health. Pp. 561-584 in *Climate Change 1995: Impacts, Adaptation and Mitigation of Climate Change: Scientific and Technical Analyses*, R.T. Watson et al., eds. Cambridge University Press, Cambridge, U.K.

McNeill, J.R. 1992. *The Mountains of the Mediterranean World: An Environmental History.* Cambridge University Press, New York.

Meltzoff, S.K., and E. LiPuma. 1986. The social and political economy of coastal zone management: Shrimp mariculture in Ecuador. *Coastal Zone Management Journal* 14(4):349-380.

Mendelsohn, R.O., and J.E. Neumann. 1998. *The Impact of Climate Change on the United States Economy.* Cambridge University Press, New York.

Mileti, D.S., T.E. Drabek, and J.E. Haas. 1975. *Human Systems in Extreme Environments: A Sociological Perspective.* University of Colorado Institute of Behavioral Science, Boulder.

Mileti, D.S., J.D. Darlington, E. Passerini, B.C. Forrest, and M.F. Myers. 1995. Toward an integration of natural hazards and sustainability. *The Environmental Professional* 17(2):117-126.

Mintzer, I.S. 1987. *A Matter of Degrees: The Potential for Controlling the Greenhouse Effect.* World Resources Institute, Washington, D.C.

Mitchell, R.B. 1994. *Intentional Oil Pollution at Sea: Environmental Policy and Treaty Compliance.* MIT Press, Cambridge, Mass.

Mitchell, R.C., and R.T. Carson. 1989. *Using Surveys to Value Public Goods: The Contingent Valuation Method.* Resources for the Future, Washington, D.C.

Moran, E.F. 1981. *Developing the Amazon.* Indiana University Press, Bloomington.

Moran, E.F. 1995. *The Comparative Analysis of Human Societies: Toward Common Standards for Data Collection and Reporting.* L. Rienner Publishers, Boulder, Colo.

Moran, E.F., and E. Brondizio. 1998. Land-use change after deforestation in Amazonia. In *People and Pixels: Linking Remote Sensing and Social Science*, D.M. Liverman et al., eds. National Academy Press, Washington, D.C.

Moran, E.F., E. Brondizio, P. Mausel, and Y. Wu. 1994. Integrating Amazonian vegetation, land use and satellite data. *BioScience* 44(5):329-338.

Moran, E.F., A. Packer, E. Brondizio, and J. Tucker. 1996. Restoration of vegetation cover in the eastern Amazon. *Ecological Economics* 18:41-54.

Morgan, M.G., and H. Dowlatabadi. 1996. Learning from integrated assessment. *Climatic Change* 34(Dec.):337-368.

Myers, N. 1993. *Ultimate Security: The Environmental Basis of Political Stability.* W.W. Norton, New York.

Nakicenovic, N. 1996. Freeing energy from carbon. *Daedalus* 125(3):95-112.

National Academy of Engineering. 1994. *The Greening of Industrial Ecosystems,* B. Allenby and D.J. Richards, eds. National Academy Press, Washington, D.C.

National Academy of Engineering. 1989. *Technology and Environment.* J. Ausubel and H. Sladovich, eds. National Academy Press, Washington, D.C.

National Acid Precipitation Assessment Program. 1991. *The Oversight Review Board of the National Acid Precipitation Assessment Program. The Experience and Legacy of NAPAP.* NAPAP, Oversight Review Board, Washington, D.C.

National Research Council. 1984a. *Energy Use: The Human Dimension,* P.C. Stern and E. Aronson, eds. Freeman: New York.

National Research Council. 1984b. *Improving Energy Demand Analysis,* P.C. Stern, ed. National Academy Press, Washington, D.C.

National Research Council. 1988. The human dimensions of global environmental change. Pp. 134-200 in *Toward an Understanding of Global Change: Initial Priorities for U.S. Contribution to the International Geosphere-Biosphere Program.* National Academy Press, Washington, D.C.

National Research Council. 1989. *Improving Risk Communication.* National Academy Press, Washington, D.C.

National Research Council. 1990. *Research Strategies for the USGCRP.* National Academy Press, Washington, D.C.

National Research Council. 1992. *Global Environmental Change: Understanding the Human Dimensions,* P.C. Stern, O.R. Young, and D. Druckman, eds. National Academy Press, Washington, D.C.

National Research Council. 1993. *Population and Land Use in Developing Countries,* C.L. Jolly and B.B. Torrey, eds. National Academy Press, Washington, D.C.

National Research Council. 1994a. *Assigning Economic Value to Natural Resources.* National Academy Press, Washington, D.C.

National Research Council. 1994b. *Science and Judgment in Risk Assessment.* National Academy Press, Washington, D.C.

National Research Council. 1994c. *Science Priorities for the Human Dimensions of Global Change.* National Academy Press, Washington, D.C.

National Research Council. 1996a. *Learning to Predict Climate Variations Associated with El Niño and the Southern Oscillation.* National Academy Press, Washington, D.C.

National Research Council. 1996b. *Understanding Risk: Informing Decisions in a Democratic Society,* P.C. Stern and H.V. Fineberg, eds. National Academy Press, Washington, D.C.

National Research Council. 1996c. *Upstream: Salmon and Society in the Pacific Northwest.* National Academy Press, Washington, D.C.

National Research Council. 1997. *Environmentally Significant Consumption: Research Issues,* P.C. Stern et al., eds. National Academy Press, Washington, D.C.

National Research Council. 1998. *People and Pixels: Linking Remote Sensing and Social Science.* D. Liverman et al., eds. National Academy Press, Washington, D.C.

National Research Council. 1999. *Making Climate Forecasts Matter.* P.C. Stern and W.E. Easterling, eds. National Academy Press, Washington, D.C.

Nelson, R.R., and S.G. Winter. 1982. *An Evolutionary Theory of Economic Change.* Harvard University Press, Cambridge, Mass.

Netting, R. 1981. *Balancing on an Alp.* Cambridge University Press, Cambridge, U.K.

Nichols, A.L. 1984. *Targeting Economic Incentives for Environmental Protection.* MIT Press, Cambridge, Mass.

Nordhaus, W.D. 1991. To slow or not to slow: The economics of the greenhouse effect. *The Economic Journal* 101(407):920-937.

Nordhaus, W.D. 1994. *Managing the Global Commons: The Economics of Climate Change.* MIT Press, Cambridge, Mass.

Nordhaus, W.D., and G.W. Yohe. 1983. Future carbon dioxide emissions from fossil fuels. Chapter 2 in *Changing Climate.* National Academy Press, Washington, D.C.

North, D.C. 1990. *Institutions, Institutional Change, and Economic Performance.* Cambridge University Press, Cambridge, U.K.

North, D.C. 1994. Constraints on institutional innovation: Transaction costs, incentive compatibility, and historical considerations. Pp. 48-70 in *Agriculture, Environment and Health: Sustainable Development in the 21st Century,* V.W. Ruttan, ed. University of Minnesota Press, Minneapolis.

Ojima, D.S., W.J. Parton, and D.S. Schimel. 1993. Modeling the effects of climatic and CO_2 changes on grassland storage of soil C. *Water, Air, and Soil Pollution* 70:643.

Ortloff, C., and A.L. Kolata. 1993. Climate and collapse: Agroecological perspectives on the decline of the Tiwanaku state. *Journal of Archaeological Science* 20:195-221.

Ostrom, E. 1990. *Governing the Commons: The Evolution of Institutions for Collective Action.* Cambridge University Press, New York.

Ostrom, E., R. Gardner, and J. Walker. 1994. *Rules, Games, and Common-Pool Resources.* University of Michigan Press, Ann Arbor.

Parry, M.L., T.R. Carter, and N.T. Konijn, eds. 1988. *The Impact of Climatic Variations on Agriculture.* Kluwer Academic Publishers, Boston.

Parson, E.A. 1993. Protecting the ozone layer: The evolution and impact of international institutions. Pp. 27-73 in *Institutions for the Earth: Sources of Effective International Environmental Protection*, P.M. Haas, R.O. Keohane, and M.A. Levy, eds. MIT Press, Cambridge, Mass.

Parson, E.A. 1996. A global climate-change policy exercise: Results of a test run. Working Paper WP-96-90. International Institute for Applied Systems Analysis, Laxenberg, Austria.

Parson, E.A. 1997. Informing global environmental policy-making: A plea for new methods of assessment and synthesis. *Environmental Modeling and Assessment* 2(4).

Parson, E.A., and K. Fisher-Vanden. 1997. Integrated assessment models of global climate change. *Annual Review of Energy and the Environment* 22:589-628.

Parton, W.J., D. Schimel, and D.S. Ojima. 1994. Environmental change in grasslands: Assessment using models. *Climatic Change* 28:111.

Patz, J.A., P.R. Epstein, T.A. Burke, and J.M. Balbus. 1996. Global climate change and emerging infectious diseases. *Journal of the American Medical Association* 275:217-223.

Peck, S.C. 1993. CO_2 emissions control comparing policy instruments. *Energy Policy* 21(March):222-230.

Peck, S.C., and T.J. Teisberg. 1992. CETA: A model for carbon emissions trajectory assessment. *The Energy Journal* 13(1):55-77.

Pielke, R.A., Jr., and C.W. Landsea. 1998. Normalized hurricane damages in the United States: 1925-1995. *Weather and Forecasting* 13:621-631.

Pielke, R.A., Jr., and R.A. Pielke, Sr. 1997. Vulnerability to hurricanes along the U.S. Atlantic and Gulf coasts: Considerations of the use of long-term forecasts. Pp. 147-184 in *Hurricanes: Climate and Socioeconomic Impacts*, H.F. Diaz and R.S. Pulwarty, eds. Springer, Berlin.

Portney, P.R. 1994. The contingent valuation debate: Why economists should care. *Journal of Economic Perspectives* 8:3-17.

Powell, W.W., and P. DiMaggio, eds. 1991. *The New Institutionalism in Organizational Analysis*. University of Chicago Press, Chicago.

Pritchett, L. 1994. Desired fertility and the impact of population policies. *Population and Development Review* 20(1):1-55.

Ramlogan, R. 1996. Environmental refugees: A review. *Environmental Conservation* 23(1):81.

Reilly, J. 1996. Agriculture in a changing climate: Impacts and adaptation. Pp. 427-467 in *Intergovernmental Panel on Climate Change. Climate Change 1995: Impacts, Adaptation and Mitigation of Climate Change: Scientific and Technical Analyses*, R.T. Watson et al., eds. Cambridge University Press, Cambridge, U.K.

Renard, Y. 1991. Institutional challenges for community-based management in the Caribbean. *Nature and Resources* 27(4):4-9.

Renn, O., T. Webler, H. Rakel, P. Dienel, and B. Johnson. 1993. Public participation in decision-making: A three-step procedure. *Policy Sciences* 26:189-214.

Renn, O., T. Webler, and P. Wiedemann. 1995. *Fairness and Competence in Citizen Participation*. Kluwer Academic Publishers, Dordrecht, Netherlands.

Richards, J.F. 1990. Land transformation. Pp. 163-178 in *The Earth as Transformed by Human Action*, B.L. Turner et al., eds. Cambridge University Press, New York.

Richardson, B.C. 1992. *The Caribbean in the Wider World, 1492-1992: A Regional Geography*. Cambridge University Press, New York.

Richels, R.G., and J. Edmonds. 1995. The economics of stabilizing atmospheric CO_2 concentrations. *Energy Policy* 23:373-378.

Riebsame, W.E. 1990. The United States Great Plains. Pp. 561-575 in *The Earth as Transformed by Human Action: Global and Regional Changes in the Biosphere Over the Past 300 Years*, B.L. Turner et al., eds. Cambridge University Press, Cambridge, U.K.

Rokeach, M. 1973. *The Nature of Human Values*. Free Press, New York.

Rosenberg, N.J., P.R. Crosson, K.D. Frederick, W.E. Easterling III, and M.S. McKenney. 1993. The MINK methodology: Background and baseline (Missouri, Iowa, Nebraska, Kansas). *Climatic Change* 24(1-2):7-22.

Rosenzweig, C. 1985. Potential CO_2 induced climate effects on North American wheat producing regions. *Climatic Change* 7:367-389.

Rosenzweig, C., and M. Parry. 1994. Potential impact of climate change on world food supply. *Nature* 367:133-138.

Rotmans, J. 1990. IMAGE: *An Integrated Model to Assess the Greenhouse Effect.* Kluwer Academic Publishers, Boston.

Rudel, T.K. 1989. Population, development, and tropical deforestation: A cross-national study. *Rural Sociology* 54:327-338

Runge, C.F. 1995. Trade, pollution, and environmental protection. Pp. 353-375 in *The Handbook of Environmental Economics*, D.W. Bromley, ed. Blackwell, Oxford, U.K.

Runge, C.F., F. Ortalo-Magne, and P. Vande Kamp. 1994. *Free Trade, Protected Environment: Balancing Trade Liberalization and Environmental Interests.* Council on Foreign Relations, New York.

Ruttan, V.W. 1997. Induced innovation, evolutionary theory, and path dependence: Sources of technical change. *Economic Journal* 107:1520-1529.

Sagoff, M. 1998. Aggregation and deliberation in valuing environmental public goods: A look beyond contingent pricing. *Ecological Economics* 24:213-230.

Sauer, C.O. 1963. *Land and Life.* University of California Press, Berkeley.

Schipper, L., and E. Martinot. 1993. Decline and rebirth: Energy demand in the former USSR. *Energy Policy* 21(9):969-975.

Schipper, L., and S. Meyers. 1993. Using scenarios to explore future energy demand in industrialized countries. *Energy Policy* 21(3):264-275.

Schipper, L., S. Meyers, R. Howarth, and R. Steiner. 1992. *Energy Efficiency and Human Activity: Past Trends, Future Prospects.* Cambridge University Press, Cambridge, U.K.

Schlager, E., and E. Ostrom. 1992. Property-rights regimes and natural resources: A conceptual analysis. *Land Economics* 68(3):249-262.

Schlesinger, W.H., ed. 1997. *Biogeochemistry: An Analysis of Global Change.* Academic Press, San Diego.

Schurr, S.H. 1984. Energy use, technological change, and production efficiency: An economic-historical interpretation. *Annual Review of Energy* 9:409-425.

Scott, F.R. 1995. *Institutions and Organizations.* Sage Publications, Thousand Oaks, Calif.

Scott, M.J. 1996. Human settlements in a changing climate: Impacts and adaptation. Pp. 399-426 in *Intergovernmental Panel on Climate Change. Climate Change 1995: Impacts, Adaptation and Mitigation of Climate Change: Scientific and Technical Analyses*, R.T. Watson et al., eds. Cambridge University Press, Cambridge, U.K.

Schwartz, S.H. 1992. Universals in the content and structure of values: Theoretical advances and empirical tests in 20 countries. *Advances in Experimental Social Psychology* 25:1-65.

Shafik, N. 1994. Economic development and environmental quality: An econometric analysis. *Oxford Economic Papers* 46:757-773.

Shavell, S. 1985. *Economic Analysis of Accident Law.* Harvard University Press, Cambridge, Mass.

Shimata I., C.B. Schaaf, L.G. Thompson, and E. Moseley-Thompson. 1991. Cultural impacts of severe drought in the prehistoric Andes. *World Archaeology* 22:247-270.

Sinclair, P.R. 1987. *State Intervention and the Newfoundland Fisheries: Essays on Fisheries Policy and Social Structure.* Avebury, Aldershot.

Sklair, L. 1991. *Sociology of the Global System.* Harvester Wheatsheaf, London.

Skole, D.L., W.H. Chomentowski, and W.A. Salas. 1994. Physical and human dimensions of deforestation in Amazonia. *BioScience* 44(5):314-322.

Slovic, P. 1987. Perception of risk. *Science* 236:280-285.

Smith, C.L. 1986. The life cycle of fisheries. *Fisheries* 11(4):20-24.

Social Learning Group. 1998. *Social Learning in the Management of Global Environmental Risks*, W.C. Clark, J. Jaeger, and J. van Eijndhoven, eds. MIT Press, Cambridge, Mass.

Socolow, R., C. Andrews, F. Berkout, and V. Thomas, eds. 1994. *Industrial Ecology and Global Change*. Cambridge University Press, Cambridge, U.K.

Solow, R. 1993. An almost practical step toward sustainability. *Resources Policy* (Sept.):162-172.

Squires, D., J. Kirkley, and C.A. Tisdell. 1995. Individual transferable quotas as a fisheries management tool. *Reviews in Fisheries Science* 3(2):141-169.

Steele, J.H. 1996. Regime shifts in fisheries management. *Fisheries Research* 25:19-23.

Stern, P.C. 1992. What psychology knows about energy conservation. *American Psychologist* 47:1224-1232.

Stern, P.C. 1997. Toward a working definition of consumption for environmental research and policy. Pp. 12-25 in *Environmentally Significant Consumption: Research Directions*. P.C. Stern et al., eds. National Academy Press, Washington, D.C.

Stern, P.C., and T. Dietz. 1994. The value basis of environmental concern. *Journal of Social Issues* 50(3):65-84.

Stern, P.C., and S. Oskamp. 1987. Managing scarce environmental resources. Pp. 1043-1088 in *Handbook of Environmental Psychology*, D. Stokols and I. Altman, eds. Wiley, New York.

Stern, P.C., E. Aronson, J.M. Darley, D. Hill, E. Hirst, W. Kempton, and T. Wilbanks. 1986. The effectiveness of incentives for residential energy conservation. *Evaluation Review* 10:147-176.

Stern, P.C., T. Dietz, and L. Kalof. 1993. Value orientations, gender, and environmental concern. *Environment and Behavior* 25(3):322-348.

Stonich, S., J. Bort, and L. Ovares. 1997. Globalization of the shrimp mariculture industry: The social justice and environmental quality implications in Central America. *Society and Natural Resources* 10(2):161-179.

Steward, J.H. 1955. *Theory of Cultural Change*. University of Illinois Press, Urbana.

Thacher, P.S. 1993. The Mediterranean: A new approach to marine pollution. Pp. 110-134 in *International Environmental Negotiation*, G. Sjostedt, ed. Sage, Newbury Park, Calif.

Thomas, W.L. 1954. *Man's Role in Changing the Face of the Earth*. University of Chicago Press, Chicago.

Tietenberg, T.H. 1985. *Emissions Trading: An Exercise in Reforming Pollution Policy*. Resources for the Future, Washington, D.C.

Tietenberg, T.H. 1991. Managing the transition: The potential role for economic policies. Pp. 187-226 in *Preserving the Global Environment: The Challenge of Shared Leadership*, J.T. Mathews, ed. Norton, New York.

Tietenberg, T.H. 1992. *Environmental and Natural Resource Economics, 3d ed.* Harper Collins, New York.

Tol, R.S.J. 1995. The damage costs of climate change: Toward more comprehensive calculations. *Environmental and Resource Economics* 5:353-374.

Toth, F.L. 1994. Models and games for long-term policy problems. Unpublished paper presented at the 1994 meeting of the International Simulation and Gaming Association, Ann Arbor, Mich.

Tucker, R.P., and J.F. Richards, eds. 1983. *Global Deforestation and the Nineteenth-Century World Economy*. Duke University Press, Durham, N.C.

Turner, B.L., II. et al. 1990. *The Earth as Transformed by Human Action: Global and Regional Changes in the Biosphere Over the Past 300 Years*. Cambridge University Press, Cambridge, U.K.

Turner, B.L., II, G. Hayden, and R.W. Kates. 1993. *Population Growth and Agricultural Change in Africa*. University Press of Florida, Gainesville.

Turner, B.L., II, W.B. Meyer, and D.L. Skole. 1994. Global land-use/land-cover change: Towards an integrated study. *Ambio* 23(1):91.

Turner, B.L., II, D. Skole, S. Sanderson, G. Fischer, L. Fresco, and R. Leemans. 1995. *Land Use and Land Cover Change: Science/Research Plan.* IGBP Report No. 35/HDP Report No. 7. Human Dimensions of Global Environmental Change Programme, Geneva.

U.S. Department of Energy. 1997. Online International Energy Databases. http://www.eia.doe.gov/fueloverview.html1#international.

U.S. Environmental Protection Agency. 1989. *Policy Options for Stabilizing Global Climate*, D.A. Lashof and D.A. Tirpak, eds. Draft Report to Congress. U.S. Environmental Protection Agency, Office of Policy, Planning, and Evaluation, Washington, D.C.

U.S. Global Change Research Prpgram. 1997. *Our Changing Planet.* USGCRP, Washington, D.C.

van Asselt, M.B.A., A.H.W. Beusen, and H.B.M. Hilderink. 1996. Uncertainty in integrated assessment: A social scientific perspective. *Environmental Modeling and Assessment* 1(June):71-90.

Vari, A., and P. Tamas, eds. 1993. *Environment and Democratic Transition: Policy and Politics in Central and Eastern Europe.* Kluwer Academic Publishers, Boston.

Vaughan, E. 1993. Individual and cultural differences in adaptation to environmental risks. *American Psychologist* 48:673-680.

Vaughan, E. 1995. The significance of socioeconomic and ethnic diversity for the risk communication process. *Risk Analysis* 15:169-180.

Vickers, D., ed. 1997. Marine resources and human societies in the North Atlantic since 1500. Paper presented at the Conference on Marine Resources and Human Societies in the North Atlantic Since 1500, Oct. 20-22, 1995. ISER Conference Paper Number 5. Institute of Social and Economic Research, Memorial University of Newfoundland, St. John's, Newfoundland, Canada.

Victor, D.G., K. Raustiala, and E.B. Skolnikoff, eds. 1997. *The Implementation and Effectiveness of International Environmental Commitments.* MIT Press, Cambridge, Mass

Viscusi, W.K., and M.J. Moore. 1989. Rates of time preference and valuations of life. *Journal of Public Economics* 38:297-317.

von Winterfeldt, D., and W. Edwards. 1986. *Decision Analysis and Behavioral Research.* Cambridge University Press, New York.

Watts, M.J., and H.G. Bohle. 1993. The space of vulnerability: The causal structure of hunger. *Progress in Human Geography* 17:43.

Wenger, D. 1985. Collective behavior and disaster research. In *Disasters: Theory and Research ,* E.L. Quarantelli, ed. Sage Publications, Beverly Hills, Calif.

Wernick, I. 1996. Consuming materials: The American way. *Technological Forecasting and Social Change* 53:111-122.

Wernick, I., and J. Ausubel. 1995. National materials flows and the environment. *Annual Review of Energy and the Environment* 20:462-492.

Weyant, J., O. Davidson, H. Dowlatabadi, J. Grubb Edmonds, E.A. Parson, R. Richels, J. Rotmans, P.R. Shukla, R.S.J. Tol, W. Cline, and S. Fankhauser. 1996. Integrated assessment of climate change: An overview and comparison of approaches and results. Pp. 367-396 in *Climate Change 1995: Economic and Social Dimensions of Climate Change.* Contribution of Working Group III to the Second Assessment Report of the Intergovernmental Panel on Climate Change, J.P. Bruce, H. Lee and E.F. Haites, eds. Cambridge University Press, Cambridge, U.K.

Wigley, T.M.L., R.G. Richels, and J.A. Edmonds. 1996. Economic and environmental choices in the stabilization of atmospheric CO_2 concentrations. *Nature* 379(Jan. 18):240-43.

Wilk, R.R. 1997. Emulation and global consumerism. Pp. 110-115 in *Environmentally Significant Consumption: Research Directions*, P.C. Stern et al., eds. National Academy Press, Washington, D.C.

Witherspoon, S., P.P. Mohler, and J.A. Harkness. 1995. *Report on Research into Environmental Attitudes and Perceptions (REAP).* European Consortium for Comparative Social Surveys.

Wood, W.B. 1994. Forced migration: Local conflicts and international dilemmas. *Annals of the Association of American Geographers* 84(4):607.

World Resources Institute. 1996. *World Resources: A Guide to the Global Environment.* Oxford University Press, New York.

Worster, D. ed. 1988. *The Ends of the Earth: Perspectives on Modern Environmental History.* Cambridge University Press, New York.

Yohe, G.W., and R. Wallace. 1996. Near term mitigation policy for global change under uncertainty: Minimizing the expected cost of meeting unknown concentration thresholds. *Environmental Modeling and Assessment* 1(June):47-57.

Yohe, G.W., J.E. Neumann, P.B. Marshall, and H. Ameden. 1996. The economic cost of greenhouse induced sea level rise for developed property in the United States. *Climatic Change* 32:387-410.

Young, M.D., and B.J. McCay. 1995. Building equity, stewardship, and resilience into market-based property rights systems. Pp. 87-102 in *Property Rights and the Environment: Social and Ecological Issues*, S. Hanna and M. Munasinghe, eds. The World Bank, Washington, D.C.

Young, O.R. 1994a. *International Governance: Protecting the Environment in a Stateless Society.* Cornell University Press, Ithaca, N.Y.

Young, O.R. 1994b. The problem of scale in human/environment relations. *Journal of Theoretical Politics* 6:429-447.

Young, O.R. 1996. Rights, rules, and responsibilities in international society. Pp. 245-264 in *Rights to Nature*, S. Hanna, ed. Island Press, Washington, D.C.

Young, O.R., and G. Osherenko, eds. 1993. *Polar Politics: Creating International Environmental Regimes.* Cornell Studies in Political Economy. Cornell University Press, Ithaca, N.Y.

Zeckhauser, R. 1975. Procedures for valuing lives. *Public Policy* 23:427-463.